T0205511

Studies in Computational Intelligence

Volume 978

Series Editor

Janusz Kacprzyk, Polish Academy of Sciences, Warsaw, Poland

The series "Studies in Computational Intelligence" (SCI) publishes new developments and advances in the various areas of computational intelligence—quickly and with a high quality. The intent is to cover the theory, applications, and design methods of computational intelligence, as embedded in the fields of engineering, computer science, physics and life sciences, as well as the methodologies behind them. The series contains monographs, lecture notes and edited volumes in computational intelligence spanning the areas of neural networks, connectionist systems, genetic algorithms, evolutionary computation, artificial intelligence, cellular automata, self-organizing systems, soft computing, fuzzy systems, and hybrid intelligent systems. Of particular value to both the contributors and the readership are the short publication timeframe and the world-wide distribution, which enable both wide and rapid dissemination of research output.

Indexed by SCOPUS, DBLP, WTI Frankfurt eG, zbMATH, SCImago.

All books published in the series are submitted for consideration in Web of Science.

More information about this series at http://www.springer.com/series/7092

Madhusudan Singh

Information Security of Intelligent Vehicles Communication

Overview, Perspectives, Challenges, and Possible Solutions

 Springer

Madhusudan Singh
Woosong University
Daejeon, Korea (Republic of)

ISSN 1860-949X ISSN 1860-9503 (electronic)
Studies in Computational Intelligence
ISBN 978-981-16-2219-9 ISBN 978-981-16-2217-5 (eBook)
https://doi.org/10.1007/978-981-16-2217-5

This Springer imprint is published by the registered company Springer Nature Singapore Pte Ltd.
The registered company address is: 152 Beach Road, #21-01/04 Gateway East, Singapore 189721,
Singapore

Preface

This book highlights are potential cyber-security overview, perspective challenges which may affect advanced Vehicular technology. It considers vehicular security issues and possible solutions, with the aim of providing secure vehicle-to-vehicle, vehicle-to-infrastructure, and inside-of-vehicle communication. This book is introducing vehicles cryptography mechanism including encryption and decryption approaches and cryptography algorithms such as symmetric and asymmetric cryptography, Hash functions and Digital Signature certificates for modern vehicles. This book will discuss cybersecurity structure and provide specific security challenges and possible solutions in Vehicular Communication such as vehicle to vehicle communication, vehicle to Infrastructure and in-vehicle communication. This book provides importance of cyber security in automotive technology by providing fruitful knowledge. It will also present key insights from security in regard to vehicles collaborative information technology cyber security whether it's time to re-thinking cyber security in advanced vehicles. The more our vehicles become intelligent, the more we need to work on safety and security for vehicles technology. This book is of interest to automotive engineers and technical managers who want to learn about security technologies, and for those with a security background who want to learn about basic security issues in modern automotive applications.

Daejeon, Korea (Republic of)   Madhusudan Singh

Contents

List of Figures

List of Tables

Chapter 1
An Overview of Automotive Vehicles and Information Security

Madhusudan Singh ⓘ

Abstract In this chapter, we present the International SAE (Society of Automotive Engineer) six level development of vehicle automation and their information security challenges. In addition, we have introduced the history of automotive technology revolution and discussed about vehicles development stages. We have discussed threats of advanced automotive technologies such as self-driving car and its components. Finally, we discuss several automotive technologies with the association of information technologies.

Keywords Intelligent vehicle · Autonomous vehicles · Information security

1.1 Introduction

The evolution of the automotive industry began in the 1890s. Now at the dawn of the twenty-first century, this industry is over 130 years old. The automotive industry is an industry characterized by a safety culture that became systemic in the 1960s. This industry is characterized by tremendous growth as the number of passenger cars alone are estimated at about one billion. Future projections are even more expansive. Unlike the automotive industry, the information technology industry started with the propagation of computers and information technology systems (IT systems) as early as the 1950s. This created a usage explosion in the 1990s that led to todays wired environment. Although this industry is only 70 years old, the processing power increased exponentially. With the combination of wireless internet and with location finding capabilities, the IT industry has created a new paradigm for human existence. The Information Technology Industry is characterized overall by a security driven culture: this stems from the fact that IT focuses on data, which analyzes trends and models [1]. In recent times, there has been an ever-increasing merging of the automotive industry with the information technology industry. Everyday vehicles utilize an

M. Singh (✉)
School of Technology Studies, Endicott College of International Studies, Woosong University, Daejeon, Republic of Korea
e-mail: msingh@wsu.ac.kr

© The Author(s), under exclusive license to Springer Nature Singapore Pte Ltd. 2021
M. Singh, *Information Security of Intelligent Vehicles Communication*,
Studies in Computational Intelligence 978,
https://doi.org/10.1007/978-981-16-2217-5_1

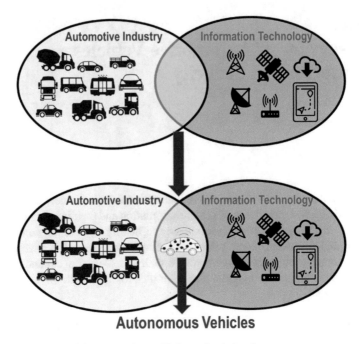

Fig. 1.1 The merging of the automotive and information industries

enormous number of electronic and communication devices that created connected the automotive safety with the need for security. As we can see in Fig. 1.1 has shown the automotive industry integrated with information technology and working to develop Intelligent vehicles. This merging has made the fantasy of an autonomous vehicle conceivable soon. Nevertheless, even in the present vehicles we get a wealth of real-time information in regard to fuel and ignition systems, power trains, brakes, transmissions, electronic and demonstrative hardware alongside parking, indicators, street risk alerts, voyage control, and route help.

Soon we are heading to complete automation that means an end-to-end automotive journey without driver intervention. In is scenario, the driver becomes the passenger! This idea was science fiction a few decades ago. The 1980s sci-fi show "Knight Rider" acquainted with the overall population the possibility of a self-sufficient vehicle named "KITT" or Knight Industries Two Thousand as shown in Fig. 1.2. The "KITT" had three modes: Normal Cruise, Auto Cruise and Pursuit [2].

- At Normal Cruise, the driver had control of the vehicle. In a crisis, the vehicle could dominate and enact Auto Cruise mode however there was a "Manual Override" to forestall this.
- At Auto Cruise, the vehicle could drive itself using a propelled Auto Collision Avoidance System.

Fig. 1.2 Knight Industries Two Thousand (KITT) autonomous car

- At Pursuit Mode, the vehicle was utilized during rapid driving circumstances and utilized a mix of manual and PC helped activity. The driver was in fact in charge of the vehicle and the PC helped direct certain moves.
- It is fascinating to take note of that in Normal Cruise, the driver had control of the vehicle yet in a crisis, the vehicle could even now dominate and actuate Auto Cruise mode indicating that innovation was better than the human information.

In any case, there was a "Manual Override" to forestall if the driver decides to do so uncovering an innate doubt to the new technology. In Fig. 1.3 has represents the connectivity of advanced vehicle of things (AVOT). It's means as much as vehicles

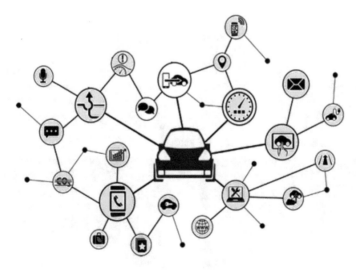

Fig. 1.3 Connectivity of advanced vehicles of things (AVOT)

are integrating with information technology as much as they are connected with things such as phone, watches navigations, service center as many as can.

The rest of the chapter is organized as follows. Section 1.2 provides an overview of automotive technology components. Section 1.3 introduces the International SAE level development of vehicle automation technology and related characteristics. Section 1.4 discusses the information security of automotive applications. Section 1.5 conclude an overview of current available automotive vehicle application and future research directions.

1.2   An Overview of Automotive Technology Components

A close examination of any current vehicle will reveal that in years that are more recent the automotive industry has been drastically augmented from mechanical and driver-only physical controls to Electronic Control Units or ECUs. There are dozens of embedded ECUs in modern vehicles that control electronic components running millions of lines of code. These units are well connected over internal buses (mainly CAN buses) to enable both critical safety and convenience features [3]. Some functionality examples include engine control, brakes, steering, phone connectivity, Bluetooth, connectivity for alerting the driver, oil pressure warnings, engine efficiency information, and voice control of car functions. Let us examine in even more detail what the merging of the automotive industry with the information technology industry has provided us so far. In Fig. 1.4 has explained the four major areas Electronics Control Units (ECUs) and other high-tech equipment controls in vehicles.

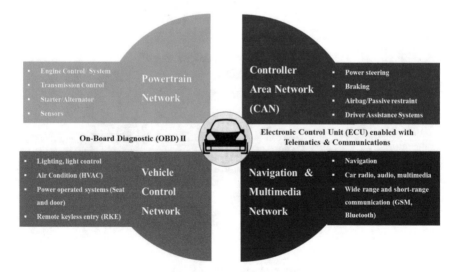

Fig. 1.4 Advanced automotive technology components

These are power train Network, the Controller Area Network (CAN), the vehicle network, and the navigator and multimedia network.

- **The Powertrain Network** permits connectivity between ECUs in the vehicle and guarantees that the engine control unit coordinates appropriately. The transmission control framework guarantees that engine torque yield is effectively moved to the street giving the traction and gives control the driver requires.
- **The Controller Area Network (CAN)** is a fast half-duplex differential and sequential communication protocol. It is utilized as multi-master system for interfacing ECUs inside the vehicle.
- **The Navigation and Multimedia Network** gives infotainment frameworks in the zones of wide range and short-range communication and navigation **The Vehicle Control Network** progresses technology and gives better coordination between vehicle and driver.
- **The On-Board Diagnostic (OBD) II** is the second generation of On-Board Self-Diagnostic. It is utilized for all sort of vehicles to give cooperation among equipment and programming in the vehicle's PC to monitor virtually every part. We will get more insights regarding OBD II later in this course. Finally, the **Electronic Control Unit (ECU)** empowered with Telematics and Communications gives the availability and information transmission between numerous buses.

1.3   Level of Vehicle Automation

SAE International, previously known as the Society of Automotive Engineers, is a U.S. based globally active professional association and standards developing organization. On your screen, you can see the updated chart they released in 2019, defining the six levels of driving automation from SAE level 0 (no automation) to SAE level 5 (full vehicle autonomy). This chart serves as the industry's most-cited reference for automated vehicle (AV) capabilities. It is predicted that by 2030 automotive technology will be fully autonomous. In Fig. 1.5 has shown the description of vehicle automation levels [4].

- *Level 0—No Automation*
  At level 0 there is no automation, and the vehicle is controlled physically by the human driver including all parts of speed and direction. Some features, for example, the emergency braking mechanism is accessible in level 0 vehicle, yet it is not automated. Figure 1.6 has presents the level 0 vehicles.
- *Level 1 Driver Assisted*: Level 1 the driver gets support for particular tasks, for example, parking, and indicators; level 1 has presented the automation feature cruise control. Another example, versatile cruise control, ensures there is a safe separation between vehicles however, the human driver monitors and controls the driving function, for example, speed, controlling, and slowing down. Figure 1.7 shown the vehicles of level 1.

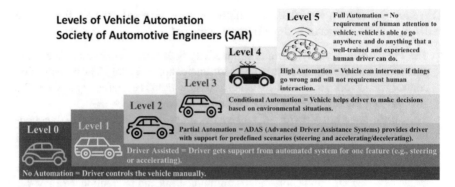

Fig. 1.5 Levels of vehicle automation

Fig. 1.6 Level 0 vehicle example

Fig. 1.7 Level 1 vehicle example

- *Level 2—Partial Automation*: Level 2 vehicles have embraced the advanced driver assistance systems (ADAS) where the driver gets some help for adapting to predefined situations because of in-vehicle sensors and different electronics that illuminate the vehicle to separate with different vehicles. Speed, focusing the vehicle on the path, and so on. ADAS can control the steering and accelerate or decelerate on the continuous moving vehicles however the human driver can assume responsibility for the vehicle whenever. Some development has been made through the journey of automation and some of them in level 2 vehicles: Tesla Autopilot, Cadillac (General Motors) Super Cruise frameworks. An overview of ADAS based vehicle has shown in Fig. 1.8.

Fig. 1.8   Level 2 ADAS enabled vehicle example

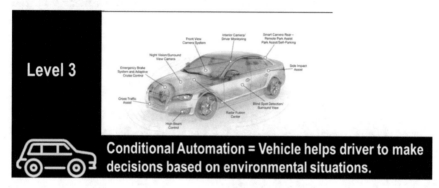

Fig. 1.9   Level 3 vehicle example

- *Level 3—Conditional Automation*: Vehicles of this level have a superior comprehension about their environment since they identify things around and make better choices. A driver can give up control to an automated system however should be prepared to reclaim control; for instance, a level 3 vehicle will decelerate in high rush hour traffic and quicken in low rush traffic. Examples of a level 3 vehicle: Audi A8L in Europe (2019) as presented in Fig. 1.9.
- *Level 4—High Automation*: Level 4 vehicles need not bother with any human mediation largely however human drivers have the alternative to drive physically. The principle key distinction between level 3 and level 4 is that level 4 vehicles can intervene on the off chance that anything unordinary ought to occur or turn out badly. Level 4 vehicles can run in self-driving mode yet because of unsupported infrastructure it works inside restricted or limited areas. Examples of level 4 vehicles: NAVYA (French organization) shuttles. Cabs in the U.S., and WAYMO self-driving taxi in Arizona (from Google) shows in Fig. 1.10.
- *Level 5—Full Automation*: At Level 5 we show up a genuine driverless vehicle. Level 5 able vehicles utilize an automated driving system (ADS) in the vehicle to monitor and move through all street conditions and require no human intercessions at all eliminating the requirement for a controlling steering wheel and pedal as well as conditional braking movement. The human occupants are simply travelers [5]. Although a significant number of the mechanical, parts exist for an artificially intelligent vehicle today, because of guidelines and legal battle in court. Level 5

Fig. 1.10 Level 4 vehicle example

Fig. 1.11 Level 5 fully automated vehicle example

vehicles are likely still a few years away (expected around 2030) and it is depicted in Fig. 1.11.

On the path to realizing a level 5 fully autonomous vehicle, several automotive technologies under development are coming soon to a parking lot near you. In the context of cyber security, we must be aware of these technologies, as they will become part of the final solution.

1.4   Information Security for Automotive Applications

Now there is a big question that whether current IT security is feasible for the future and current Automotive Technology and applications. Indeed, future automotive technology and applications will have a vital need for IT protection. Below are some advantages of IT security with reference to the embedded automotive systems. Some applications domains are listed in Fig. 1.12, where IT Security can play an effective and efficient role for Automotive Technology.

- IT security will enhance the reliability of the control systems in the automotive and can increase the fault tolerance of the Telematics systems.
- IT security can play a very efficient role in dealing with the novel innovations and technologies introduced in the Automotive.

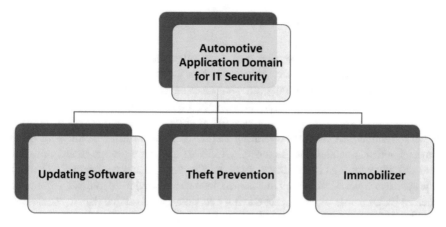

Fig. 1.12 Automotive application domain for IT security

- **Updating Software**: There is a need to update the Electronic Control Units (ECU) in the car because after the car is shipped many software errors are found in them. Secondly, many users want to configure their car according to their utility, comfort and price. There is some non-profiting updating for the manufacturers, where the owner updates a feature, which looks attractive, and with no cost payment. Embedded Security technologies such as Digital signature, encryption can restrict the user to feely update the software's according to his will.
- **Theft Prevention**: The electronic security device immobilizer was used way back to prevent is maybe the oldest type of IT security in Automotive and it has benefited by avoiding car thefts up to 60% over the last 15 years. Similar cryptography methods such as Identification protocols and tamper resistance protocols can be proposed for each part of the automotive to avoid them from being stolen or illegal exchange.
- **Immobilizer**: An immobilizer or immobilizer is an electronic security gadget fitted to an engine vehicle that halts the engine from running except if the right transponder key (or other tokens) is available. This keeps the vehicle from being "hotwired" after entry has been accomplished and hence reduces engine vehicle burglary. Examination shows that the uniform utilization of immobilizers diminished the pace of vehicle burglary by 40%.

Automotive systems are executed as a network of embedded smart devices, some of which have worldwide availability and established communication. Conventional vehicle security arrangements, similar to caution, keyless section, and so forth neglect to ensure the automotive IT security framework.

1.5 Automotive Vehicle Applications

New technologies need to connect with each other and bond seamlessly with the people utilizing them. Let us examine how this takes place in the next few paragraphs. A Connected car uses integrated information technology that is well equipped with a wireless network that communicates with other telecommunication systems. For example, a connected car will use the smartphone operating system and the telecommunications network operating standards to share information with other devices both inside and outside the vehicle. Next, there are groups of technologies that provide warnings to the driver. Advanced driver-assistance systems (ADAS) is one such example. ADAS will synchronize the data received from multiple resources, such as ECUs, cameras, and make driving decisions based on that data to accelerate, decelerate brake, etc.

Other technology is the Connected HD; these vehicles use data transmission within ECU CAN and sensors (such as protocol stacks, object detection, Ethernet and security image analysis, graphic processing, traffic signal recognition, etc.) to make decision so the vehicle drives smoothly on road. Autonomous vehicles have technology strategically positioned. This image identifies where each area of technology is positioned [6].

- **LIDAR and Radar Safety Sensors**

LIDAR estimates separation by enlightening targets with pulsed laser light and measure reflected pulses with sensors to make a 3-D guide of the region.

Cameras give real-time vehicle sideways collision i.e., obstacle recognition to encourage path takeoff and track street information (like street signs). Radar sensors monitor the situation of the vehicles close by. The Dedicated Short-Range Communications (DSRC) device license a vehicle to establish connections and communicate with different vehicles (V2V) utilizing a wireless communication standard that empowers reliable and trustworthy information transmission in active safety applications. The Central Computer gets data from different parts and coordinates vehicle largely. Ultrasonic Sensors utilize high-frequency sound waves generated from an attached sensor in the vehicle framework that bobs back to compute separation between vehicles on the road and obstacles. GPS draws the pattern of the vehicle utilizing satellites and wholly designed in triangular pattern. Either current GPS technology is constrained to a specific separation vehicles or obstacles, yet propelled GPS is being developed. Figure 1.13 has demonstrated the case of innovation situating in innovative vehicles design. The position can change the spot as per manufacturing organization.

- **Google Autonomous Vehicle**

In addition to the technology in an autonomous vehicle, each vehicle will have different sensors positioned to assist in maneuvering the vehicle. This image is a 3D rendering of an autonomous self-driving electric car using LIDAR and Radar Safety sensors. Notice the three Emergency Braking sensors in the front of the vehicle, the

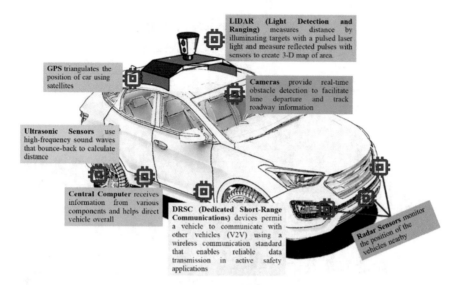

Fig. 1.13 An example of technology positioning in vehicles

three Parking Aid System sensors in the rear bumper of the vehicle, and the four Blind Spot Detection sensors on the sides of the vehicle that will assist in switching lanes [7].

In Fig. 1.14 has shown a working example of a Google's self-driving car. On the top of the vehicle, there are sensors such as lasers, radars and cameras that detect other cars and objects in all directions. The rounded shape of the vehicle is rounded not so much for aerodynamic reasons but to allow for maximum sensor field of view [8]. The interior is designed with the premise that everybody is a passenger meaning it is designed for riding not driving [9]. The on-board computer is quite unique in the sense that it has been designed and built for the sole purpose to accommodate self-driving. In the Google self-driving car, battery banks provide the power as the vehicle is 100% electric. Lastly, the back-up systems provide redundant systems for steering and braking.

A well-established standard is consistently an assurance that procedures and usage are agreeable with best practices and rules. A few endeavors have just been attempted to give such cyber security rules to the car business, universally yet additionally nation explicit. The covering hazard is a reality, and all these normalization bodies need to arrange with one another to maintain a strategic distance from clashes and ambiguities. Nearby activities incorporate for example the EVITA (E-Safety Vehicle Intrusion Protected Applications) venture in Europe, which planned to give in-vehicle reference engineering dependent on HSM. The Japanese IPA (Information Promotion Agency) vehicle data security direct secured a start to finish lie-pattern of the vehicle including outsider and providers conduct toward security. Even more as of late, universal normalization bodies, for example, ISO (International Organization for Standardization) and SAE (Society of Automotive Engineers) joint their push to

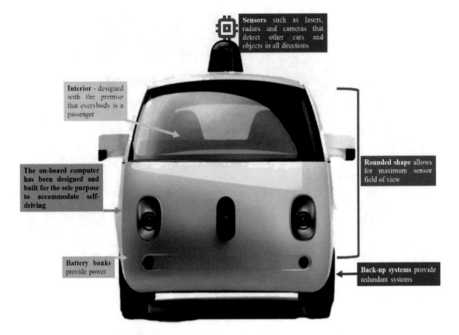

Fig. 1.14 Example: Google autonomous vehicle [8]

deal with the meaning of a devoted cyber security standard for the car business. SAE as of now started these works inside the Vehicle Electrical System Security Committee where the J3061 cyber security manual and the J3101 prerequisites for equipment ensured security archives are delivered. ISO's TC22 and SAE are likewise recognizing the possible communications between framework wellbeing and cyber security.

1.6 Conclusion

In this chapter, presents the history of automotive technology that is included in six layers. Which begins with layer 0 it has manual vehicle to Layer 5 that is full automation. The advanced components of vehicles is On-Board. Electronic Control Unit etc. has described with example of Self-driving car. With the huge changes that will go with the presentation of ADS in an expanding scale, it very well may be foreseen that recognizable impacts will occur on user experience, traffic wellbeing, effectiveness, portability, efficiency, energy, environment, and economy. It is additionally obvious that ADS will cause unpleasant problematic changes in certain businesses, and the arrangement way may not be as convenient and smooth as certain defenders may want. Building up an all-around grounded and methodical assessment system and making reasonable and down to earth devices to examine the social advantages of ADS stay a difficult yet commendable theme for future projects.

Acknowledgements This research was funded by Woosong University Academic Research in 2021.

References

1. G. Ur-Rehman, A. Ghani, M. Zubair, S.H.A. Naqvi, S. Muhammad, D. Singh, IPS: incentive and punishment scheme for omitting selfishness on the internet of vehicles (loV). IEEE Access (2019). https://doi.org/10.1109/ACCESS20192933873
2. Knight Rider (2008 film), https://en.wikipedia.org/wiki?curid147467112008
3. The 5 Levels of Autonomous Vehicles (2018), https://www.truecar.com/blog/5-levels-autono mous-vehicles/
4. SAE J3016, Levels of Driving Automation (2019), https://www.sae.org/news/2019/01/sae-upd ates-j3016-automated-driving-graphic
5. Sunil Raj Thota Blog, Wanna go for a long ride cars. Let's Kickstart the 'Driving Innova-tion—Self-driving'. https://sunitrajthota.blogspot.com/201904wanna-go-for-long-ride-lets-kic kstart-the-driving-innovation-self-driving-carshtml (2019)
6. R.P. Prakash, R. Tripathi, D. Singh, Analytical model for clustered vehicular ad hoc network analysis. ICT Express (2018). https://doi.org/10.1016/j.icte.2018.01.001
7. D. Singh, G. Tripathi, S.C. Shah, R. da Rosa Righi, Cyber-physical surveillance system for the internet of vehicles, in *JEEE World Forum on Internet of Things WF-IoT 2018*, Singapore, 5–8 Feb 2018, pp. 551–555
8. M. Singh, IEEE E-Learning, Evolution of intelligent and autonomous vehicles (2020). https:// ieeexplore.ieee.org/courses/details/EDP585
9. D. Singh, M. Singh, Internet of vehicles for smart and safe driving, in *2015 International Conference on Connected Vehicles and Expo (ICCVE)*, Shenzhen, China, 19–23 Oct 2015, pp. 328–329 (2015)

Chapter 2
Security Analysis of Information Technology

Madhusudan Singhⓘ

Abstract As we know security is always important challenges in technology. In cyberworld is full of lack of security. We need to protect our cyber worlds in real time. This article has described basics of security, security impact and provide and an overview on security areas and their vulnerable places in our cyber world such as Computer security, Network Security, and Cyber security. These security features must not be limited to protecting confidential information, but also needs to address safety critical systems such as brake, accelerator or steering etc.

Keywords Security · Cyber security intelligent vehicle · Autonomous vehicles · Information security

2.1 Introduction

Security is protected the assets of the system from malicious vulnerability and mitigate their impact on the system. The assets can be possible any object or entity or system of any organization or organization itself. Any assets can be exploited by attacker/hacker for their own benefits due existing vulnerability in the system. The attackers/hackers exploit the system with the help of system vulnerability and, it has made possible to illegally access the system or modify any asset of the system. The malicious users (Attacker/Hacker/Unauthorized) can find a single week point as individual or as group in a system or framework and targeting that single to harm or damage or illegally access the whole system information or that system itself [1]. In Fig. 2.1, we can see the security requires in every area, digital (e.g., Internet), physical (e.g., vehicle), and non-physical (e.g., Wireless environment).

The security resources are itself very important assets of any system that should be protect the system against malicious threats and attacks. The environment of

M. Singh (✉)
School of Technology Studies, Endicott College of International Studies, Woosong University, Daejeon, Republic of Korea
e-mail: msingh@wsu.ac.kr

© The Author(s), under exclusive license to Springer Nature Singapore Pte Ltd. 2021   15
M. Singh, *Information Security of Intelligent Vehicles Communication*,
Studies in Computational Intelligence 978,
https://doi.org/10.1007/978-981-16-2217-5_2

Fig. 2.1 Overview of security

the system and environment are impact each other such as if security assets are environment friendly then it has added on for system environment and vice versa.

The rest of the chapter is organized as follows. Section 2.2 discussed the concept of security; Sect. 2.3 gives the overview of computer security with the network security. Section 2.4 presents vulnerabilities and attacks in computer/IT Security. Section 2.5 describe the details of security goals and final Sect. 2.6 has conclude security details.

2.2 Security Conceptual

There are certain conceptual definitions of security which comes across now and then, whenever security is referred in different security areas [2]. Figure 2.2 has shown the different way of security concept.

- *Security Objective*: A statement stating high level organization's security goals and business security needs to achieve.
- *Constraint*: A restriction that provide hindrance so that the security objectives can be achieved.
- *Security Mechanism*: A mechanism that detects, prevent, and mitigate the harm done by an attack.
- *Threats*: A harm or danger or potential loss to the assets.
- *Vulnerability*: A fault, flaw or weakness which can be exploited by a threat.

Fig. 2.2 Security conceptual

- **Defense in Depth**: An absolute range of multiple layers of security protection measures for the overall security of the system.
- **Risk**: An event that could compromise the critical assets of the system.
- **Policy**: Security rules which put constraints on the users of the system to abide by these rules for the protection of the system.
- **Countermeasure**: Security measures used to mitigate threats to achieve the security objective.
- **Assurance**: A confident declaration that the security mechanism will meet the expectation.
- **Resilience**: The potential ability to retrieve swiftly from harm, attacks, and threats.

2.3   Overview of Computer Security

The computer security ensure that peripherals of computer system are protected from outsider and achieved the tried protection features which are CIA (Confidentiality, Integrity, Availability). A computer has basic components which are hardware, software, firmware, and data which are termed as the main assets in any computer system.

2.3.1   Computer Protection

There are malicious users, intruders and hackers who hack, steal, and modify your personal and critical assets. Assets can be any important data, information, bank money or anything which is important to you. We can protect our assets by following simple but very useful steps. In Fig. 2.3, we can see ways of the computer security.

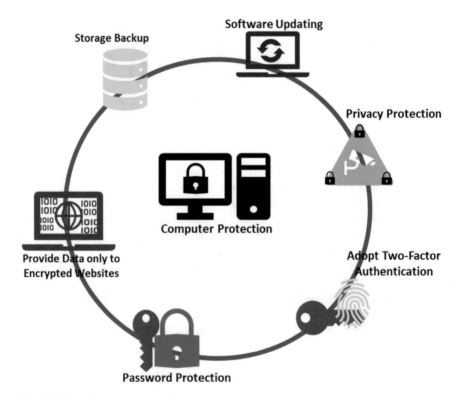

Fig. 2.3 Ways of computer protection

They are described as follows:

- **Software Updating**

Systems software's, Application software's and programming software's should be refreshed routinely, particularly the internet browsers that are utilized for associating the Internet on the grounds that doing so can shield them from more up to date and lately developed attacks and threats that infected your software's and in the long run your PC. It is imperative to introduce great observing applications, which can identify the infectious virus and alleviate the issues it caused.

Software Updating

- **Privacy Protection**

It is not wise to trust any email, webpage, phone call or message and share your private information. There are many spams and phishing activities which steal, and hack your important assets, such as your important card numbers, bank account, and security numbers and passwords. Be aware and know of such witty scammers as they act very trustworthy and reliable.

Privacy Protection

- **Password Protection**

We may have come across such login sites which do not accept an only alphabet, or only numbers or only alpha numeric or only special character passwords. They accept your password which has numbers, alphabets, and special characters all together. This combined password is termed to be very strong and not easy to guess or hack.

Password Protection

Below are mentioned some ways to make your passwords strong enough to keep them safe from hackers.

- Try to form long passwords combining numbers, alphabets, and special characters, not less than 10 characters. You may also use Capital letters and small letters together.
- Try not to make obvious passwords which resemble your name, date of birth, dates or ordered numbers, instead try something unique combination of numbers, alphabets, and special characters.
- Try not to use similar password for many of your accounts. If your password is stolen, then all your accounts are at risk.
- Don't ever share your passwords by texting, emailing or via voice.
- If you are liable to forget your password, then be cautious to keep your written password safely out of reach of everyone.

- **Adopt Two-Factor Authentication**

As the name suggests Two Factor Authentication utilizes two factors for authenticating your identity. One factor is your password, and the other factor may be a simple code sent to your mobile phone registered with your account or any random alphabet or number generated by an application [3]. So, even if your password is compromised, the second factor can be saving your account from hacking.

**Adopt Two-Factor
Authentication**

- **Provide Data Only to Encrypted Websites**

During internet shopping, use only encrypted websites. These sites will encrypt the personal information in your computer before sending it to the website's server. The websites which use encryption use Hyper Text Transfer Protocol Secure (HTTPS) before the web address [4]. This is different from the commonly used HTTP as it uses an additional S which stands for secure.

**Provide Data only to
Encrypted Websites**

- **Storage Backup**

Never re-lie on single storage of your important files. Always make a backup of your important files to other storage mediums. Even if your computer is compromised, you can still have access to your files via alternate storage [5].

Storage Backup

2.3.2   Overview of Network Security

Network security is a vast topic and is an area of specialization to the Information Technology (IT). Internet users are getting aware of the importance of Security and its role. In the recent years a huge number of internet users are exploring security associated websites. The Banking applications ask to install security plugins or programs before any transaction of money. The security certifications have gain popularity. Biometric security measures such as fingerprint and retina identification which were once read only in science fictions have become very common and adopted by a lot of gadgets for secure authentication. Although, the awareness and importance of security has increased, still many organizations when developing a technology or a gadget; give a secondary thought to security after development being the primary [6]. They don't have a well sought out plan for security from the very initial phase of elicitation of requirements and planning. Computer security ranges from protecting

Fig. 2.4 Network security overview

the hardware, software and the information broken into bits. Figure 2.4 has shown the overview of network security.

Next, we describe the vulnerabilities in the network and the importance of network security and later, we will discuss the methods to overcome them.

- **Wired and Wireless Physical Networks**

In a wired network, as the names suggests that computers are connected through wires; wire can be copper wire, twisted pair or fiber optic. Wired networks work on the principles of Ethernet protocol. Computers are first connected to wires or the Unshielded Twisted Pair (UTP) cables, which in turn are connected to multiple switches and which further are connected to the router for internet connectivity. In Fig. 2.5 has represents the wired network.

In Fig. 2.6 has shown the wireless network.

In contrast the wireless network as the name suggests, the computers are not connected with wire instead they are connected to access points via radio transmissions [6]. Further, the access points are connected via cables to switch/router for internet connectivity.

Although both wired and wireless networks are used in organizations the wireless networks are more popular than wired networks because of the factor of mobility. Advantage of wireless networks is that multiple devices can be used remotely and can share files and resources remotely. Wireless networks increase accessibility and ease of use.

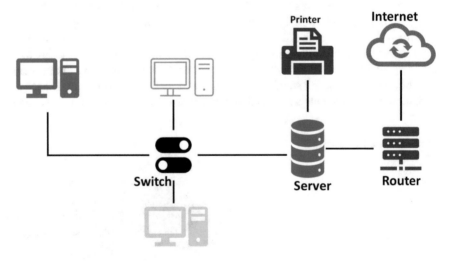

Fig. 2.5 Wired network

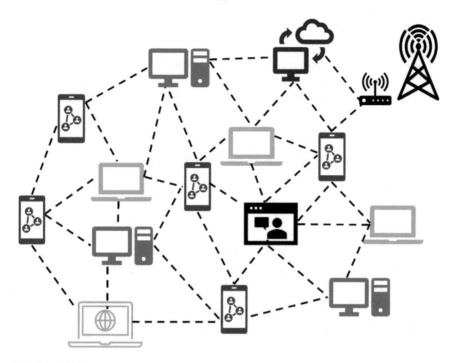

Fig. 2.6 Wireless networks

2.4 Vulnerabilities and Attacks

If we talk about vulnerabilities and attacks in terms of wired and wireless networks, then attacker can be easily attack on wireless networks due it's unsecure switch port and to attack on switch port attackers doesn't require any essential physical connection for device [7]. The most common vulnerability in wired and wireless networks is accessing the network unauthorized way. At below, we can find some other vulnerabilities that can place after unauthorized.

- Attackers' attacks during the data transmission in the networks and sniffing the data packet.
- Attackers can transmit the bogus information in the network and overloaded with unauthentic data.
- Attacker placed himself between receiver and sender nodes within the network and the MAC addresses of authentic hosts are spoofed to capture data and induce Man-in-the-middle attack.

Therefore, we cannot take security as non-essential elements in networks. Security is very important elements for the system, and we must be think beginning of any system designing and implementation to remove vulnerabilities from the network.

2.5 Security Goals

It is essential to have successful primitives to guarantee the security of the nodes in a vehicular network. All in all, tending to the security of any system requires the accompanying security administrations: confidentiality, integrity assurance, availability, authentication, and non-repudiation. The automobile system also requires having these security services in order to provide safe and appropriate operation of the system. Figure 2.7 shows the security goals.

As shown in Fig. 2.7 security goals are discussed below:

2.5.1 Confidentiality

A confidentiality mechanism guarantees the secrecy of the transmitted information by guaranteeing that the message isn't unveiled to an unapproved client/user. Confidentiality can be accomplished by utilizing cryptographic primitives that gives encryption/decryption functionalities. These type of cryptographic primitives transforms the message in such a way that only legitimate nodes (e.g., sender and receiver) can comprehend the actual meaning of the message while it seems meaningless to the other nodes.

Fig. 2.7 Security goals

2.5.2   Integrity Assurance

Integrity assurance (often also referred as data integrity) of a message provides the receiver with an assurance that the data has not been modified during transmission. Data integrity affirmation can be accomplished by utilizing a Message Authentication Code (MAC) tag. Note that this system doesn't keep an assailant from altering the message, rather it gives the collector an intend to identify unapproved alteration of the message.

2.5.3   Availability

Availability ensures information assets such as session key and applications are accessible by the authorized users. Cryptographic techniques cannot provide availability. Rather availability can be ensured by having proper backups, redundant units/paths etc.

2.5.4   User Authentication

User authentication is the process of verifying something to be true. Particularly a system needs to verify the ID, location, and property of the sender. In general authentication is the mechanism that verifies the claimed identity of a user. This guarantees the received message is really from the sender who he/she is professing to be. Cryptographic procedures, for example, the digital signature can be utilized to give client/user verification. These are systems where one party (sender) can sign the message utilizing a digital signature, while the other party (collector) can confirm the message is really from the authorized sender or not.

2.5.5   Non-repudiation

Non-repudiation ensures that the network entities (sender and receiver) cannot falsely deny a prior communication.

2.5.6   Network Security Goals

As discussed so far, we realize that there are huge weakness in the any network even data transmission within network does not reliable it's vulnerable with attacks. Firstly,

Fig. 2.8 Network security
goals

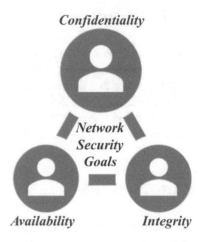

attacker transmits an enormously data in network and based on that data it's acquiring
the communication channel and accessing the real transmits data and misused the
network data. Figure 2.8 has represents the security goals of the networks.

The network security provides the security for the entire network and end to end
connectivity. It does not target to secure the end devices only.

Network security has numerous secondary goals, for example, reliability,
usability, integrity, and wellbeing of data and the network [8]. The essential goal of
system security is the groups of three Security which are Confidentiality, Integrity,
and Availability as shown in Fig. 2.8.

- *Confidentiality*: The point is to secure the critical assets (information) from unapproved clients in confidentiality. The confidentiality for security guarantees that
  the critical elements are open just to approved clients.
- *Integrity*: This aims to ensure that the data is not modified or tampered by
  unauthorized users during its transmission in the network.
- *Availability*: The aim of availability to assure that availability of resources and
  data are available, whenever it requested by the legal and authorized users.

2.6 Summary

We studied at the beginning of this chapter that Security is the protection of critical assets of the system. There are some common terms such as Cybersecurity,
IT Security, Network Security, and Computer Security. All their aim is protection
of computer systems, their hardware, software, and their data from damage, harm
or attack from unauthorized users. The cybersecurity is extending and emerging
essential part of every level in the technology such as supercomputer to smallest
communication device or technology.

Acknowledgements This research was funded by Woosong University Academic Research in 2021.

References

1. A. Alshnoul, Information systems security measures and countermeasures: protecting organizational assets from malicious attacks. Commun. IBIMA **2010**, 1–9. Article ID 486878 (2010). IBIMA Publishing
2. B.B. Naik, D. Singh, A.B. Samaddar, H.-J. Lee, Security attacks on information-centric networking for healthcare system, in *19th International Conference on Advanced Communication Technology (ICACT)*, South Korea, 19–22 Feb 2017, pp. 436–441
3. J. Browning, *White Paper: Protecting Your Assets: Information Security in the Teradata Database* (TERDATA, US, 2015)
4. R. Zhuang, A.G. Bardas, S.A. DeLoach, X. Ou, A theory of cyber attacks, in *Proceedings if the Second ACM Workshop on Moving Target Defense (MTD-15)* (2015)
5. N. Ahmad, M.K. Habib, Analysis of network security threats and vulnerabilities by development & implementation of a security network monitoring solution, Master thesis, Department of Telecommunication, Blekinge Institute of Technology, Sweden, 2016
6. M. Singh, S.-G. Lee, W.K. Tan, J.H. Lam, Throughput analysis of wireless mesh network testbed, in *International Conference on Convergence and Hybrid Information Technology (ICHIT-2011)*. CCIS, vol. 206, Daejeon, Korea (2011), pp. 54–61. https://doi.org/10.1007/978-3-642-24106-2_8
7. J. Thomas, Tutorials, Omnisecu, Primary Goals of Network Security-Confidentiality, Integrity and Availability (2017), https://www.omnisecu.com/ccna-security/primary-goals-of-network-security.php
8. M. Singh, Secure ID-based routing data communication in IoT. EAI Endorsed Trans. Internet Things **18**(6) (2018). ISSN 2424-1399. https://doi.org/10.4108/eai.15-1-2018.153566

Chapter 3
Cybersecurity in Automotive Technology

Madhusudan Singh

Abstract We're in the twenty-first century, and we're no longer encircled by mechanical vehicles, but instead, there's COW (Computer on Wheels). Various security devices, gadgets and protocols for vehicles, for example, programmed emergency braking, forward-crash alarms, vehicle-to-vehicle communication, and potentially, in the coming years, completely automated vehicles, can be actualized because of the cutting-edge technologies as of now existing. Remembering that above mentioned advancements have extraordinary potential in them, vehicle producers and transportation specialists are improving the protection of technologies. This chapter will help to get a concise outline of the Importance of cybersecurity in automotive technology.

Keywords Intelligent vehicle · Autonomous vehicles · Information security

3.1 Introduction

Automotive are systems which are supported by hardware and software. Every new vehicle product is coming with new IT features year by year which are leading the production in terms of advancement of vehicle, cost, and technology. The cost of IT electronics is almost half of total of all manufacturing costs. Now a days, more than 90% of vehicles innovation are in hardware and software of vehicles [1]. The task of hardware in vehicles has listed below, which are controlling and directs the vehicles to perform the tasks.

- The essential hardware of the system that calls primary systems of the vehicles. Such as engine, advanced driver assistant system, electric system, brake system, and dashboard.
- The next secondary important elements in a vehicle are as following, Ignition, Indicators, Window control, wipers, light.

M. Singh (✉)
School of Technology Studies, Endicott College of International Studies, Woosong University, Daejeon, Republic of Korea
e-mail: msingh@wsu.ac.kr

© The Author(s), under exclusive license to Springer Nature Singapore Pte Ltd. 2021   29
M. Singh, *Information Security of Intelligent Vehicles Communication*,
Studies in Computational Intelligence 978,
https://doi.org/10.1007/978-981-16-2217-5_3

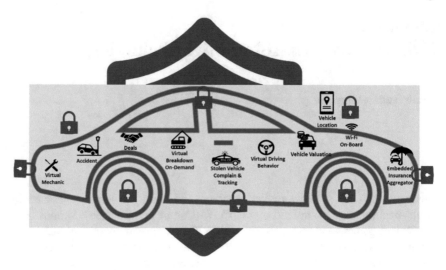

Fig. 3.1 Automotive cyber security

- The 3rd elements are Infotainment applications, such as, Navigation route systems, Telematics, back seat entertainment, music and video diversion, and GPS based administrations.

Now a days, implication of electronics devices in vehicle have significantly increasing day by day [2]. These kinds of vehicles are not completely mechanical after the mediation of electronics in them.

Present-day progressions in electronics have significantly changed the automotive industry. These vehicles are no more totally mechanical system after the mediation of electronics in them. Furthermore, these electronic devices have included a lot of unfathomable features, improving the general execution of the vehicle. Figure 3.1 has represented the overview of automotive cyber security.

Vehicles are snappier, further developed, and all the more impressive today. These headways are the aftereffect of many ECUs (electronic control units) and immense network connectivity connecting them and empowering an entirely different driving experience: from vehicles that can be remotely locked or opened to vehicles that can be controlled without an ignition key and can even drive or park themselves [3]. You can locate Google's driverless vehicles driving in Nevada USA, and they are additionally permitted to drive in Florida and California. So, it's just a matter of to what extent until we see increasingly autonomous and smart vehicles all around, most likely improving driving safety—however, shouldn't something be said about the security perspectives? Not exclusively are new vehicles fitted with these cutting-edge technologies, yet more established vehicles despite everything have a portion of these features and design.

Vehicle manufacturers contend not just on the equipment format of the vehicle, or on the quality and execution of the engine, but mostly on the new features; the driver is offered. This "new driving experience" is accomplished by utilizing

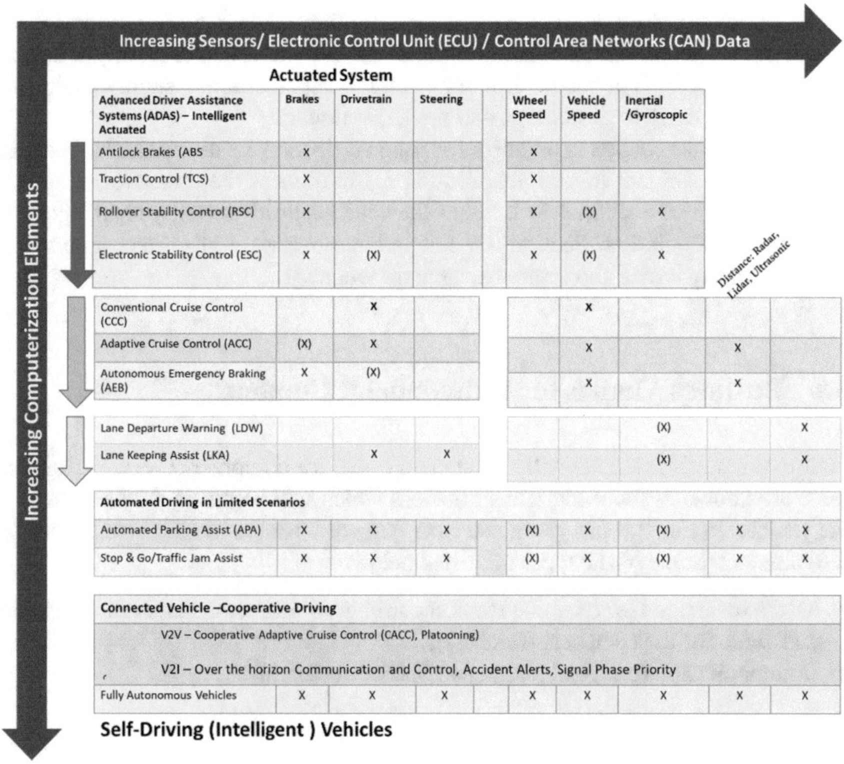

Fig. 3.2 Development of intelligent vehicles

many megabytes of code in the ECUs of the vehicle. Figure 3.2 has shown the transformation components of development of automotive technology.

The term cybersecurity is known to quite some people in this century. Over the past few decades, the introduction of computers, Internet, Satellites, and similar technologies have totally revolutionized our lives [4]. As these systems are becoming an integral part of our lives, the possibility of attacks on these systems is increasing alongside. Cybersecurity came into knowing, so as to protect these technical systems from any kind of cyberattacks. There is a need of applying cybersecurity principles at various components of the vehicles to govern safety and protect from any kind of evil attacks, damage, illegitimate access, or something which may compromise with safety of the vehicle or the driver.

The push for innovative upgrade and Internet network in all parts of present-day life is profoundly changing our utilization of devices and equipment that was not planned with this usefulness. In fact, technological advancements are changing the automotive industry by expanding the number of segments inside the vehicle that are constrained by computers and which give data to a remote system by different specialized strategies. These progressions make new open doors for cybercrime, especially demonstrations of cyber trespass by the network hackers. Specialized threats might be constrained using criminological hypothesis, explicitly a normal exercise viewpoint. The job of different actors in the automotive supply chain, and their likely capacity to discourage and moderate threats.

This rest of the chapter is organized as follows. Section 3.2 discuss about business model of infotainment in vehicles. Section 3.3 introduces the attack techniques in automotive world. Section 3.4 discusses the vehicles inside security challenges and possibilities of solution. Section 3.5 conclude an overview of current automotive vehicle security issues and future research directions.

3.2 Business Models for Infotainment Content

Future Automotive vehicles will be equipped with some equipment for entertainment; these are known as the infotainment devices which will be put in the dashboard of the vehicle. In Fig. 3.3 has given use clear view of business models component for infotainment. Some of the types are listed below:

- Media diversion (for example Book recordings, digital broadcasts, radio, music, and video for the back seat watchers),
- Communication (e.g., cell phone and SMS, Bluetooth),

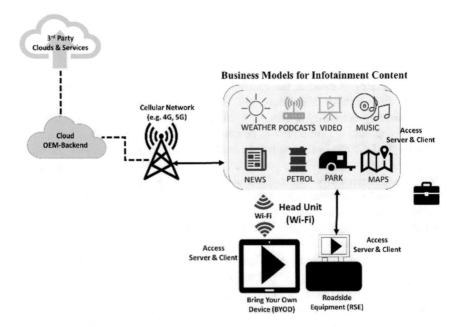

Fig. 3.3 Business models component for infotainment system

- Information system (e.g., route framework warnings of traffic data, vehicle updates).

The Infotainment content providers have a lot of opportunity to create advanced Infotainment business models in the media entertainment, Communication, and Information Systems. Here security comes into picture where the business partners can cheat by replicating the content in an illegal way. Some measures can be taken as mentioned below:

- Strong security is required for Communication system in the car so that there is secure data transmission between the caller in the car and the attendant in the outside environment.
- To avoid unauthorized copying, Digital Rights Management (DRM) is needed.
- The customers' data collection must be done in a way to secure the privacy of the users.
- To avoid improver handling of security mechanism, secure hardware parts are needed.

3.2.1 Personalization of Cars

A lot of car functions can be personalized such as notifying the users in case of traffic delays on their route and booking a restaurant, reserve parking space, social connectivity etc. Figure 3.4 has shown the elements of personalization of the cars.

Authentication of the legitimate driver is also very important factor, and it can be done by using token-based approaches such as two factors verifying from cell

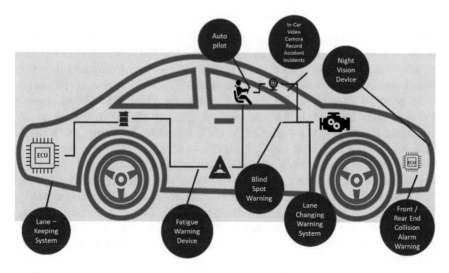

Fig. 3.4 Elements of personalization of cars

phones, smart cards. Other biometric approaches such as fingerprint recognition and Iris recognition can also be used.

3.2.2 Access Control for Car Data

Event Data recorders record a lot of data about the automobile components and the driving behavior of the driver. Cars are getting manufactured with these Event Data recorders. It is very important to access control the car data and driver's driving behavior. Access control can be achieved by using authentication, identification, and communication security mechanism.

- Anonymity: The IT technologies built-in the automotive can ignore the Privacy privileges of the driver. Circumstances, for example, event data recording which is clarified above and different situations when the driver demands for navigation information or data of disrupting any traffic norm can be a danger that can be maintained by a strategic distance from by incorporating access control technology and anonymization.
- Legal Obligations: There are some legal guidelines that require IT security, for example, Highway Toll fees. Later on, there will be more applications, for example, event data recorders, crisis calling, and so forth will require IT security because of legal commitments. Later on, embedded security and IT security will assume a key job in future automotive services, highlights and applications.

3.3 AUTOSAR

AUTOSAR (AUTomotive Open System Architecture) was created in 2003 and the primary objective was to build up an infrastructure that is autonomous of ECU hardware and diminishes the refinement of software in connected vehicles [5]. 80% of overall worldwide production is centered around AUTOSAR and now it is proclaimed as the accepted standard for the automotive software. There is an abstract layer of hardware shaped via AUTOSAR or more its applications are composed and are self-reliant from the first supplier of the ECU hardware. Protection protocols are indicated by the AUTOSAR standard that can be utilized by software modules and consolidated in the vehicle system. AUTOSAR interfaces and techniques have safe on-board network, and the specific application is controlled by OEMs who pick the cryptographic methods and encryption strategies to be utilized and applied in the vehicle system. Crypto Service Manager (CSM), Crypto Abstraction Library (CAL) and Secure On-Board Communication (SecOC) are the three principal security mechanisms in the AUTOSAR. Figure 3.5 have shown the complete overview of AutoSar.

Fig. 3.5 AUTOSAR overview

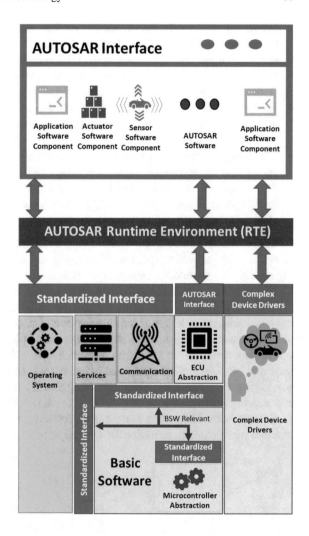

3.4 Concerns in Automotive IT Community

Below are listed some advantages of using Embedded IT security in cars:

- Enhanced Reliability
- Protection from alteration from unauthorized, owners and maintenance staff
- Introduce novel IT-based business models.

In-order to build up strong embedded security solutions, below listed are some concerns which are needed to overcome the existing problems:

- Almost every automotive application will require IT security in the coming future.
- Many IT based business models for automotive will require IT security in future.

- Integrated Security in the embedded devices shall be invisible. Security will be invisible incorporated into built-in devices. The manufacturers need to gain expertise in the embedded security technologies area.
- There is a need of at most attention while designing Security solutions because a single bug in the system's design can make the complete solution insecure.
- Embedded security in vehicles is constraint to specific considerations such as processes with restrained memory and computation, physical security and constrained cost requirement.
- It is significant to know the actual designer of the security architecture and who is controlling the cryptographic keys to dodge security structure entailments from the multiple client creation chains for autonomous and modernized vehicles.
- Many epic applications will develop in the wake of consolidating the automotive IT and the built-in security network.

The connected vehicle idea isn't being driven exclusively by advancements in automotive innovation, yet these improvements are critical to its encouraging. Figure 3.6 has shown the Automotive IT elements.

It is essential to think about the connected vehicle as an incorporated system, and as an associated element, possibly interfacing with V2I, V2V, V2IoT, and its own inner automotive system, turning into a piece of a greater associated 'ecosystem' that could conceivably include those particular application advancements. Every one of these innovations has been considered with at least one helpful destination. intelligent vehicle re-routing around blocked areas of a town or city, for example, would help ease roads turned parking lots, give drivers guidance ahead of time of approaching postponements, or give the information to empower them to embrace an elective method of getting to their goal. As they grow, such advancements would likewise make chances to utilize the current street transport system and discover a

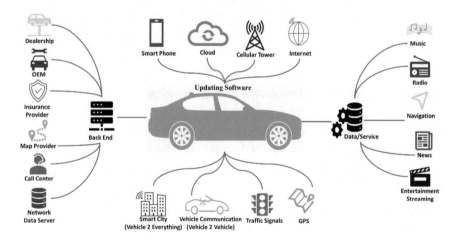

Fig. 3.6 Automotive IT elements

few answers for street usage issues that may some way or another have brought about an exorbitant and argumentative new vehicle infrastructure.

3.5 The Rising Threat in Automotive World

As we discussed in previous section, now days vehicles are highly equipped by electronics and computers to control the decisive operations, such as acceleration, brakes, stability control, and engineer performance, air bag functionality. The vehicles are becoming more safer and secure with electronic control units (ECUs) makes more complex and also increase the vulnerability in the vehicles. The ECUs are in the vehicles are interconnected with Control Area Network (CAN) bus in an unsecure fashion.

The CAN bus has introduced in 80s without any security guidelines, which sent the messages with an ID include the length (DLC) and the payload itself therefore, it's giving the receiver of the message no means to know to message sender and message legality. The messages are also inferred from the message IDs therefore allowing any component connected to the bus ability to send high priority messages, flooding the bus and "silencing other components" [6].

Nowadays, cybersecurity problems are getting more and worse. Remote hacking and immobilization of vehicles can occur. Vehicles can be hacked via the wireless air pressure sensors in the tires and can be physically harmed if someone gets hold of the CAN bus through one of the vehicle's external interfaces. Even higher cars like the Jaguar suffered from defects, causing "blue screen of death," immobilizing them.

Extremely vulnerable security architecture with many potent computers having wireless entryway (GPS, RDS cellular, IR, Wi-Fi, etc.) is seen in the Vulnerability assessments of several cars. Taking control over such a computer can lead to full control over messages which are being sent over the bus, thus entirely compromising the vehicle. As shown in Fig. 3.7.

We recently witnessed a rising wave of cyber-attacks, striking targets all over the world. If we realize that these are computers that we are driving, there is no reason to think that these computers will be left out of the game for too long. One may wonder why anyone would want to hack a vehicle. The motivation may vary; from car theft and stealing personal data to extortion, damaging business competitor's reputation or even assassination or terrorism.

3.6 Automotive Cybersecurity Overview

Research on CAN bus security has as of late expanded, for the most part, because of various exhibitions of the vulnerability of the established in-vehicle network. Koscher et al. were the first to lead ongoing attacks on automobiles. Utilizing the

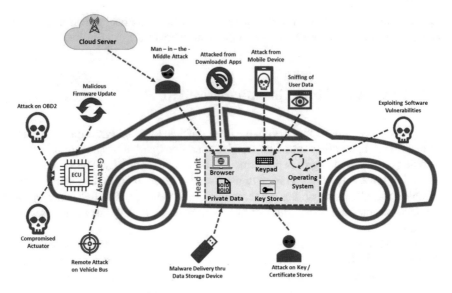

Fig. 3.7 The rising threat in vehicles

CAN bus arrange sniffing, fuzzing, and figuring out of the ECU code, they were fit for controlling various automotive features, for example, halting the engine, deactivating the brakes [7], etc. Later Checkoway et al. have demonstrated that a vehicle can likewise be utilized wirelessly, with no sort of physical access to the vehicle, utilizing a variety of techniques, for example, Bluetooth, cell radio, and even TPMS (Tire Pressure Monitoring System) [8]. The attacks on Ford Excape and Toyota Prius were effectively completed by Valasek and Miller by means of the CAN bus network design [9]. They messed with the speedometer, guiding, braking, route system, and a few other electronic hardware. In 2015, it came in notice that they had remotely impaired the braking mechanism of a Jeep while driving, which caused Chrysler to review in excess of a million vehicles. Foster and Koscher had sketched out the potential vulnerabilities in the new commercial OBD-II dongles which support cellular communication, which can even be abused through SMS. A later hack was distributed by Keen Security Lab on a Tesla model, in which scientist has control of the vehicle by misusing the bug in the infotainment's browser, which forced the company to release an over the air software update. Another recommended way to deal with secure the CAN bus was to utilize authentication of the messages on the bus by utilizing the MAC. Another idea as recommended by VAN Herrewege et al. was to employ the utilization of new light-weight protocol which can fit the CAN bus limitations better (Fig. 3.8).

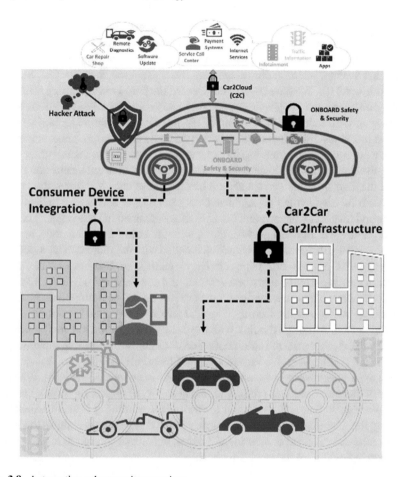

Fig. 3.8 Automotive cybersecurity overview

3.7 Automotive Cyber Security Challenges

Due to the nature of the automotive industry, automotive cybersecurity poses a non-trivial challenge on security experts. The number of companies and manufacturers participating in the stages of cars' production is enormous, so one must assume that at least some of them have vulnerabilities in their components. Since any unsecured link in the final product makes the product secure, one can infer that the cars are inevitably vulnerable.

Similar challenges might be found in the aerospace, railways and marine domains as well as in Industrial Control Systems (ICS).

As part of our research, we have thoroughly compromised a car infotainment system using a standard external interface. Hacking an ECU or any other embedded equipment is not that different from hacking a PC based system. In our lecture we will shortly refer to our experience of hacking into a vehicle's component: We have decided to check ourselves how simple it is to hack a vehicle's electronic component, specifically an infotainment system. We purchased a widespread after-market navigation system, which includes external interfaces such as radio, CD, SD card, Bluetooth, etc. [10]. We have reverse engineered the hardware, including schematic, sniffing the traces, etc. We have extracted the binary code from the system and started reversing it, with focus on the parsing of the input media from external interfaces. We have found several vulnerabilities which we could exploit for code execution and started working on implementing the attack. We wanted some visual effect on the system itself, which will indicate the fact that the system is under our control. We figured the display of the system should be the path to take. We found the function responsible for displaying the graphical background layer and used it to display our background instead of the original backgrounds contained in the system. The background picture was injected as data, along with the buffer overflow code, and after some processing, in memory, this function is called with the address in memory of the picture that was inserted. From this activity, we came know that, it was rather simple to hack an electronic component of a vehicle. And people with knowledge of reverse engineering can do this task easily and in very less time. The bottom line: Cars were designed for safety, not for security. Physical attack on a vehicle is possible and not too complicated. Although the link with the outside world allows several innovative technologies, it still opens the automobile and its systems to a possible remote attack. Attacks on Jeep Cherokee, Nissan electric car, Chevrolet Corvette, and Tesla S model are real examples of some successful cyberattacks in connected vehicles. While the volume of these automobiles progresses, the necessity of protecting them would also grow. We believe that it is important to resolve these security issues immediately. Utilizing threat modeling techniques already in the design process allows to identify safety faults early instead of later identification, which can contribute to the recall of several vehicles still on the road (Fig. 3.9).

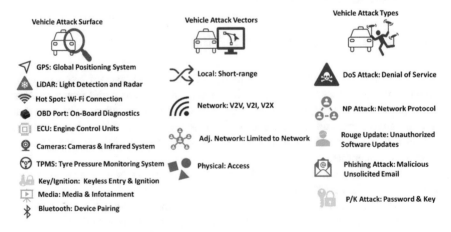

Fig. 3.9 Automotive cyber security challenges

3.8   Automotive Cybersecurity Issues in Industry

Subsequent to leading a progressively careful review of related published methodologies, we chose two threat modeling models transcendent in the computing industry (TARA and STRIDE) that appeared to be undeniably adjusted to the automotive industry.

- **Threat Agent Risk Assessment (TARA)**

The Threat Agent Risk Assessment (TARA) [11] approach has been created by security experts from Intel Security and is centered around three classes of information accumulated, alluded to as libraries. We can see all three classes in Fig. 3.10

- Threat Agent Library (TAL)—TAL library comprises of significant threat operators with their related features.

Fig. 3.10 Threat agent risk assessment (TARA) classes

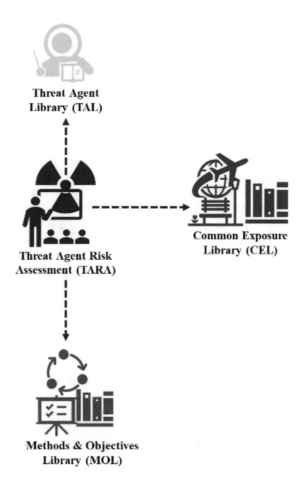

- Methods and Objectives Library (MOL)—MOL library comprises of operation in which each threat agent may apply by a related effect level.
- Common Exposure Library (CEL)—CEL library comprises of fields of the most generous exposure and powerlessness.

The security experts centrally occupy these libraries inside the company. These libraries are created by listing breach reports, incidents reports, security measures and other confidential information. Security experts can cite information from libraries and can determine the threat agent feature is needed by the threat agent to impact the danger in the corporation and the resources of the entire organization. Most recent methods and most disclosed field lists can also be known from the library. Each of the fields revealed is represented with the frequency of exposure, existing security controls to secure this field and suggested security controls. The field which requires an amended security measure is measured by taking the difference in actual and desired protection measures. Security experts can decide the areas which are more vulnerable to attacks, what operation and by which threat agent, by simply aggregating information from all the three libraries. These libraries act as an important database, which provides all the necessary information to coordinate a security policy and identify the most important vulnerabilities. Such repositories may be modified and used in subsequent situations using the TARA system.

- STRIDE

The STRIDE [12] technique was initially made by Microsoft. This method bestows perceptivity into potential attack situations by perceiving threats in the design period of any software or hardware. There are two kinds of STRIDE methods: per-cooperation and per-component. The security expert initially makes Data Flow Diagrams (DFD) of the system for analyzing and utilizing the strategy. The DFDs exhibit the pattern of interaction between the components under analysis. At that point, the methodology investigates these diagrams to distinguish likely threat to the structure. Threats are arranged into six different areas: Spoofing, Tampering, Repudiation, Disclosure of Information, Denial of Service, and Elevation of Privileges. The assessment of DFD diagrams is done physically or by utilizing the Microsoft Threat demonstrating tool which utilizes the STRIDE per-interaction variant. We can see the overview of Stride challenges and security services for stride attacks. Figure 3.11 has presented STRIDE security services relationship map.

3.9   Attack Techniques in Vehicular Environment

The vehicular ad hoc network (VANET) is essentially a wireless network and may be accessed by adversaries to mount different attacks. Here we discuss some of the attack techniques on VANET with respect to the breach of different security services.

Fig. 3.11 Stride security service

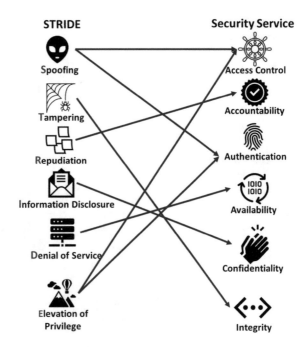

3.9.1 Attacks Related to Availability

Some of the common attacks related to security goals of availability of vehicular network resources are described as below.

- **Denial of service attack**: The attacker expects to keep the approved clients from getting to the correspondence channel or system assets. Ordinarily, this could be achieved by utilizing the sticking or flooding technique. In the sticking strategy, the attacker meddles the communication channel utilizing signs of the same frequency of the communication channel. In the flooding strategy, the attacker floods the communication channel by sending an enormous number of fake messages. This will bring about high clog in the network which may keep the real justified nodes from getting to the required systems networking assets. As example shown in Fig. 3.12.
- **Broadcast tampering attack**: The assailant transmits counterfeit messages in the system to make misperception about the right security messages among the genuine nodes. This is for the most part an insider attack as shown in Fig. 3.13.
- **Black hole attack**: In black hole attack the malevolent nodes present themselves inside a course between the sender and beneficiary that has the briefest distance/hop count. This pulls in the sender to send the packet by means of the malicious node, and once the packet is sent by means of that way, the malicious node drops those packets. In the black hole attack, the attacker drops any packet that sends via the malicious node as shown in Fig. 3.14.

Fig. 3.12 Denial of service
in vehicular communication
between car and network
server

Denial of service attack

Fig. 3.13 Broadcast
tempering attack

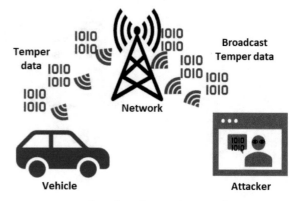

Broadcast Tempering Attack

Fig. 3.14 An example of
broadcast tempering attack

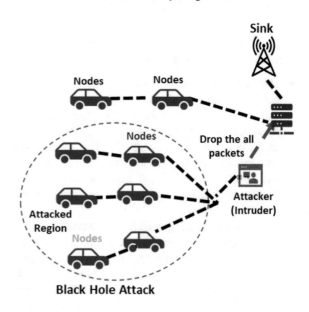

Black Hole Attack

Fig. 3.15 Grey hole attack

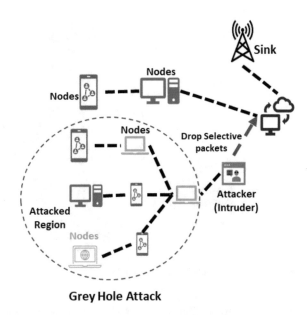

Grey Hole Attack

- *Gray hole attack*: This attack is like the black hole attack, yet on this occasion the malicious node doesn't drop all the packets that are being transmitted, rather it just performs specific packet dropping dependent on specific prerequisite as shown in Fig. 3.15.

3.9.2 Attacks Related to Confidentiality

Some of the common attacks related to security goals of confidentiality of vehicular network are described as below.

- *Eavesdropping attack*: This attack aims to obtain confidential information by listening to the communication channel. Eavesdropping can result in unauthorized disclosure of messaged and could be significantly devastating. Particularly if a system transmits its communication in plaintext over an insecure channel then attacker may steal any information just by observing the communication over the channel.
- *Traffic analysis attack*: This attack aims to deduce information by intercepting the messages over a communication channel and then analyzing the pattern of the messages. This type of attack can even be performed for encrypted communications such as shown in Fig. 3.16.

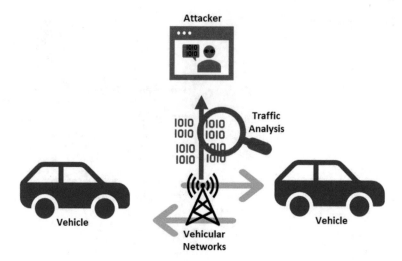

Fig. 3.16 Traffic analysis attack example

3.9.3 Attacks Related to Integrity Assurance and Authentication

Some of the common attacks related to security goals of integrity assurance and user authentication of vehicular network are described as below.

- **Message tampering attack**: In this assault, the attacker alters the message imparted between two parties. The altering may occur in three distinct structures: message suppression, message modification, message creation. In the message suppression attack, the attacker selectively removes some messages which may contain critical information. In the message alteration attack, the attacker alters some part of the message. In the message fabrication attack, the attacker creates a new false message pretending to from someone else.

  As we can see, in Fig. 3.17, Vehicle 'A' has sent hello message to vehicle 'B' but attacker illegally received vehicle 'A' broadcasted message and tempered the message and send to vehicle 'B'. When Vehicle 'B' received the message from attacker, it has believed message sent by vehicle 'A' and vehicle 'B' respond the tempered message accordingly.

- **Replay attack**: In the replay attack, the attacker copies the legitimate messages that were transmitted at a previous time, and then retransmits those messages back on the network again. Figure 3.18 has represented an example the replay attack.

- **Sybil attack**: This attack tries to forge fake identities to interrupt the usual behavior of the vehicular network. An attacker sends multiple messages with multiple identities and broadcasts its various positions at the same time. This will create confusion in the network.

  As we can see in Fig. 3.19 White color vehicle are fake vehicle, that's make

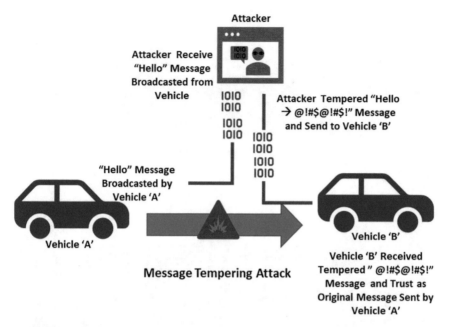

Message Tempering Attack

Fig. 3.17 An example of message tempering attack

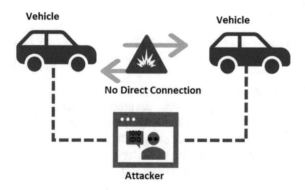

Fig. 3.18 Replay attack example

confusion between real vehicles that traffic has high congestion.

- **Impersonation attack**: The attacker introduces itself as an approved node in the network by professing to be some other specific node. The objective of this attack is either to have unauthorized access to the network assets or to upset the ordinary progression of the network. Normally these sorts of attacks are performed by taking identity credentials.

  As we can see Fig. 3.20, an accident message has broadcasted in the network but when emergency vehicle (Police) tries to communicate with vehicle (Blue

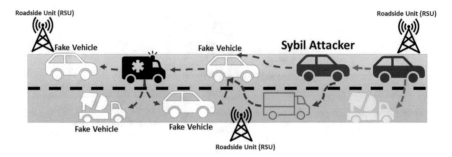

Fig. 3.19 Sybil attack

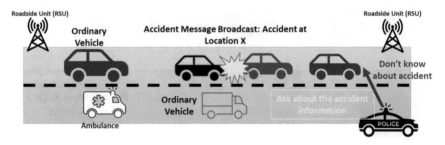

Fig. 3.20 An example of impersonation attack

Color). It has tried update the message ("Don't know about accident") and claims that message comes from original authenticated source (Accidental vehicle on Location X).

- **Vehicular Ransomware Attack**

  This article addresses the requirements and procedures for carrying out a vehicular ransomware attack [9]. In Fig. 3.21 has shown the vehicle an example of vehicle elements where ransomware attack can possible.

- **Vehicle Ransomware Attack Prerequisites**

  The cybercriminal will require in any event the accompanying attributes to do a vehicular ransomware attack:

  – Ransomware malware user and server program for the on-board acknowledgment of cyber theft on the objective vehicle alongside the proper remote control.

  – A mysterious botnet for universal conveyance and remote control of ransomware vehicle users.

  – In-vehicle protection commonly misuses a ransomware malware user alongside Trojan programming to enter and corrupt a connected in-vehicle computer.

  – An on-board lock or block activity for a basic vehicle part that cannot (effectively) be reestablished, evaded, or that cannot manage the cost of a protracted

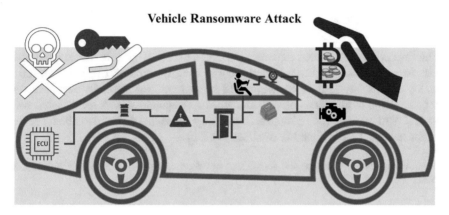

Fig. 3.21 An example of vehicle ransomware attack

time of disappointment; obviously joined with a (mystery) unlock direction to release locked vehicle segment once the ransom has been paid.
– A mysterious payment system to gather ransom and to make sure about the blackmailer from exposure and inevitable legitimate activity. Embedded Security Advancements in Vehicles.

3.10 Embedded Security and Information Technology (IT) Security

Inserted Security was presented during the 1990s. Embedded Security has gazed upward as cryptographic changes and its development or security designing in the Security community. The sorts of security issues found in embedded security are unique in relation to computer networks. The embedded security manages issues which are relating to the security of the hardware and software while IT Security manages security arrangements relating to systems and information transmission, for example, firewalls, Intrusion discovery, and intrusion detection, and so forth. There are some similarities and a number of differences between IT Security and Embedded security, some are as follows:

• Embedded Devices are small microcontrollers of 8 bit, or 16-bit size and they are very low in computation, power and memory utility. On the other hand, Personal Computers, are high power computational devices with huge so they do not confine the cryptographic usage.
• The potential attackers of embedded devices are within the reach of the device whereas Computer devices can be hacked of harmed remotely.
• Embedded systems are very cheap and cost sensitive and it is not wise to secure them with costly security solutions. On the other hand, Personal computers and laptops are comparatively more costly and it is legitimate to spend on them for cost feasible security solutions.

3.11 Summary

This report has introduced the automotive cybersecurity and the threats and also discussed possible ongoing automotive cyber security solutions overview. The criticality of this issue might be heightened by patterns, for example, exceptionally robotized and autonomous driving and V2X correspondences. A levelheaded way to deal with security can be founded on a comprehension of hazard—a blend of the seriousness and the probability of effective attacks.

Acknowledgements This research was funded by Woosong University Academic Research in 2021.

References

1. C. Paar, Embedded IT security in automotive application—an emerging area, in *Embedded Security in Cars*, ed. by K. Lemke, C. Paar, M. Wolf (Springer, Berlin, Heidelberg, 2006)
2. M. Singh, A. Singh, An empirical study on automotive cyber attacks, in *The 4th IEEE World Forum on the Internet of Things (WF-IoT 2018)*, Singapore, 05–08 Feb 2018
3. A. Palanca, E. Evenchick, F. Maggi, S. Zanero, A stealth, selective, link-layer denial-of-service attack against automotive networks, in *Detection of Intrusions and Malware, and Vulnerability Assessment. DIMVA 2017*, ed. by M. Polychronakis, M. Meier. Lecture Notes in Computer Science, vol. 10327 (Springer, Cham, 2017)
4. M. Markovitz, A. Wool, Field classification, modeling and anomaly detection in unknown CAN bus networks. Veh. Commun. **9**, 43–52. ISSN 2214 - 2096 (2017). https://doi.org/10.1016/j.vehcom.2017.02.005
5. B. Zou, M. Gao, X. Cui, Research on information security framework of intelligent connected vehicle, in *ICCSP'17: Proceedings of the 2017 International Conference on Cryptography, Security, and Privacy* (2017), pp. 91–95. https://doi.org/10.1145/3058060.3058064
6. J. Pelzl, M. Wolf, T. Wollinger, Automotive embedded systems applications and platform embedded security requirements, in *Secure Smart Embedded Devices, Platforms and Applications*, ed. by K. Markantonakis, K. Mayes (Springer, New York, 2014)
7. R. Cassettari, L. Fanucci, G. Boccini, A new hardware implementation of the advanced encryption standard algorithm for automotive applications, in *2014 10th Conference on Ph.D. Research in Microelectronics and Electronics (PRIME)*, Grenoble (2014), pp. 1–4. https://doi.org/10.1109/PRIME.2014.6872672
8. A. Karahasanovic, P. Kelberger, M. Almgren, Adapting threat modeling methods for the automotive industry, in *15th Embedded Symposium of CAR [ESCAR] Conference*, Berlin (2017)
9. M. Wolf, R. Lambert, A.-D. Schmidt, T. Enderle, WANNADRIVE? Feasible attack paths and effective protection against ransomware in modern vehicles, in *15th Embedded Symposium of CAR [ESCAR] Conference*, Berlin (2017)
10. M. Singh, D. Singh, A. Jara, Secure cloud networks for connected & automated vehicles, in *2015 International Conference on Connected Vehicles and Expo (ICCVE)*, pp. 330–335, Shenzhen, China, 19–23 Oct 2015. https://doi.org/10.1109/ICCVE.2015.94
11. A. Bolovinou, A. Ugur, A.T. Sheik, O. Ur-Rehman, G. Wallraf, A. Amditis, TARA+: Controllability-aware Threat Analysis and Risk Assessment for L3 Automated Driving Systems (2019). https://doi.org/10.13140/RG.2.2.23901.67044
12. G. Macher, E. Armengaud, E. Brenner, C. Kreiner, Threat and risk assessment methodologies in the automotive domain, Procedia Computer Science, **83**, 1288–1294 (2016), ISSN 1877-0509. https://doi.org/10.1016/j.procs.2016.04.268

Chapter 4
An Assessment of Automotive Cyber Security

Madhusudan Singh ⓘ

Abstract As we probably are aware these days, automotive technology is a developing exploration territory for scientists and researchers all over the world. The developers are taking a shot at to make astute and self-driving vehicles in, and transportation system in automotive innovation. The more vehicles become shrewd the more we have to think and work on wellbeing and security for automotive and vehicle innovation. This presentation intends to encourage conversation on automotive cybersecurity-related issues and their potential solutions.

Keywords Automotive cyber security life cycle · Layers of cybersecurity in vehicles · Cyber security threats

4.1 Automotive Cyber Security Lifecycle

The Automotive Secure Development Lifecycle involves seven-step process with a complex combination of multiple actions that act as layers of cyber security. The seven-step process includes System Design Architecture, Asset Protection Definition, Threat Model, Defining Counter Measures, Best Practices, Security Assessment, and Incident Response plan. Each Process must begin from the very beginning, at the design process and be implemented during the time of manufacturing [1]. Figure 4.1 has shown the complete life cycle of Automotive Cyber Security.

The seven-step process of automotive security development lifecycle has explained. The System Design Architecture provides advice on the efficient use of security technologies and management of the associated risks. The Asset Protection Definition has to do with anti-theft, revenues streams, that provide protection of software enabled functionality. The threat model focuses on spoofing identity, tempering data, repudiation, information disclosure, denial of service, etc. When we discuss counter measures, we are dealing with restrictions to reach certain features

M. Singh (✉)
School of Technology Studies, Endicott College of International Studies, Woosong University, Daejeon, Republic of Korea
e-mail: msingh@wsu.ac.kr

© The Author(s), under exclusive license to Springer Nature Singapore Pte Ltd. 2021
M. Singh, *Information Security of Intelligent Vehicles Communication*,
Studies in Computational Intelligence 978,
https://doi.org/10.1007/978-981-16-2217-5_4

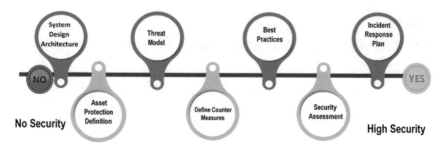

Fig. 4.1 Automotive security development lifecycle

and binding occurs only with appropriate network interfaces. **Best practices** refer to proper security awareness training and specific technology or service selections regarding security and privacy advice. **Security Assessment** involves the examination of penetration and attack testing, runtime analysis and fuzz testing, and finally static source code analysis. The **Incident Response Plan** involves three responses if a breach has been made: publicly releasing details regarding one or more security flaws, malware created by hackers and released in the wild and firmware installed on the telematics module by a hacker [1].

The rest of the chapter is organized as follows. Section 4.2 provides an overview of automotive cyber security levels and layers in vehicular environment. Section 4.3 discusses the automotive cyber security challenges and possible solutions. Section 4.4 autonomous vehicles cyber security risks. Section 4.5 conclude an overview of current available automotive vehicle cyber security challenges and possible solutions with future research directions.

4.2 Levels of Automotive Cyber Security

The automotive security can be protected from attackers in 3 level: **The design level**, the build level and the service level. We need to address security issues in each of these levels separately. At the design level, engineers are designing the vehicle hardware and software with protection features of cyber security in mind. At **the build level**, which is during the actual construction of vehicle, the manufacturer must consider the vehicle communication threat model and provide solutions to as many threats as possible. At **the Service Level**, engineers ensure that the architecture of autonomous vehicles has all security features. Vehicles must be able to control vulnerabilities of pre/post threats to allow users to use the services with minimum or very low risk. Figure 4.2 has explained the three level of automotive cybersecurity [2].

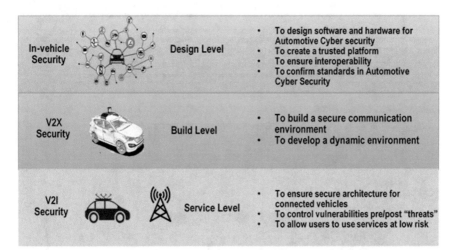

Fig. 4.2 The 3 level of automotive cybersecurity

4.2.1 Automotive Cyber Security Layers

Automotive cybersecurity can divide the whole cyber security process in 3 layers. Below, in Fig. 4.3 has show the top to bottom security overview within 3 layer. Where, **Layer 1** represents In-Vehicle Security or the protective sensor data. In this layer each vehicle needs to protect sensor data such as the Electronic Control Unit or ECU, the Control Area Network or CAN, and On-board Devices or OBDs.

In **Layer 2** insurance is given to every single vehicular communication, for example, V2V (Vehicle to Vehicle), V2I (Vehicle to Infrastructure), and V2P (Vehicle to Person).

Finally, in **Layer 3** protection is provided to the security for the data and services exchanged within the Intelligent Transportation System infrastructure security.

From below mention schematic it becomes clear that automotive technology needs an end-to-end security approach in order to defend against a myriad of stakeholders with thousands of vulnerable attack points that exist in all three layers of the network.

4.3   Automotive Cyber Security: Challenges and Possible Solutions

Automotive technology faces cyber security challenges in three main areas: Technical Safety, Technical IT Security, and Quality Attributes [3]. As shown in Fig. 4.4.

In the **Technical Safety** category protection is made against technical failures such as engine failure, light failure, tire failure, fire, brake failure, etc. These types of

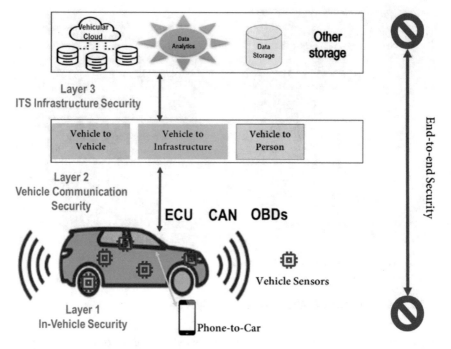

Fig. 4.3 Automotive cyber security layers

failures can be avoided with better ECU, CAN, and coordination inside the vehicle. CAN updates the messages in time to help of sensor coordination.

In the **Technical IT Security** category protection is made against malicious IT attacks that can break privacy, manipulate information, and create data intrusion. These type attacks can be avoided by managing safety and security during the software development or upgradation.

Lastly, in the **Quality Attributes** category the biggest challenge is to ensure end-user trust on applications and interconnected computing platforms and security-relevant organizational policies and processes. The solution is to ensure the trustworthy open automotive ecosystem and application store from the end users perspective.

4.4 Autonomous Vehicle Cyber Security Risks

In this section, we list the five most common cyber security risks faced by autonomous vehicles [4]:

- *Automotive Attack Vectors*: In this type attackers try make confused or force or block the communicater e.g. driver/vehicles that received messages are correct and

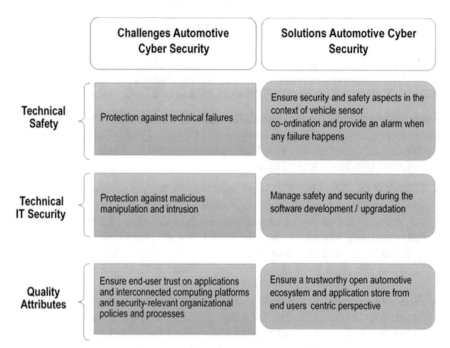

Fig. 4.4   Automotive cyber security—challenges and possible solutions

it's not harmful for network or to them. The Man-in Middle, Side Channel, Software modifications, Denial of Service etc. attacks are some example of automotive attack vector.

- *Automotive Attack Goals*: The most of attackers goals to gain the safety critical functions of vehicles and theft the personal information of vehicles/vehicles driver.
- *Automotive Vulnerable Systems*: The most vulnerable systems in vehicles are driving strategy (GPS/Navigation), vehicle control, barking acceleration, and most of connect services.
- *Automotive Immediate Risks*: Any Vehicles immediate risks of attacks are compromised vehicles, any vehicles safety functions, misbehavior of vehicles in critical situation, and many more [5].
- *Long-Term Automotive Risks*: The long term of attack risk of any vehicles has dramatic loss of consumers trust into new vehicles features, declining the attractiveness and trust by consumers into the vehicle's brands without clear cybersecurity strategy.

4.5 Automotive Cyber Attacks Types, Threats

If automotive cyber-attacks are classifying the automotive cyber-attacks into types and threats categories. So it can be category in four main types: **Bogus Information Attacks, Denial of Service (DoS), Masquerading and Data Manipulation** [6].

Bogus Information Attacks: Bogus information attacks are attacks where the hacker transmits bogus information in the vehicular environment that influences the driver's decisions for the purpose of trying to maliciously disturb the whole network. Bogus Information attacks shows in Fig. 4.5.

- *Denial of Service (DoS)*: Denial of Service or DoS are attacks where the hacker tries to create artificial network congestion by broadcasting too many messages and jamming the communication channels [7] as shown in Fig. 4.6.
- *Masquerading Automotive Attacks*: In this type attacks, attacker gain the access to vehicular network and "masquerade's" or pretends to be some else by using fake identities and pursue malicious objectives. Figure 4.7 has shown as an example of Masquerading Automotive attacks [8].
- *Data Manipulation Attacks*: Data Manipulation attacks are attacks where the hacker tries to access the data from the vehicular cloud and manipulate the data. These types of manipulated data attacks destroy the integrity of the vehicular network and cause other vehicles to make erroneous decisions based on manipulated data. Fig. 4.8 has represented an example of data manipulation attacks [9].

Fig. 4.5 Bogus information attacks

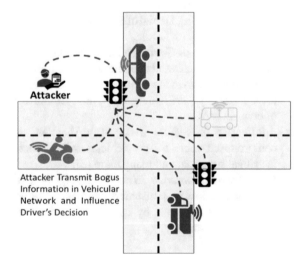

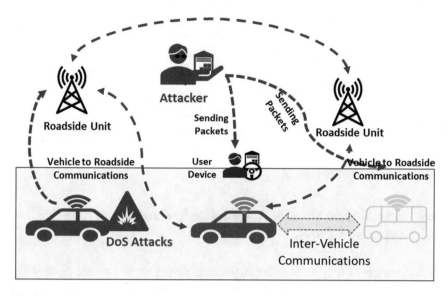

Fig. 4.6 Denial of service in automotive attack

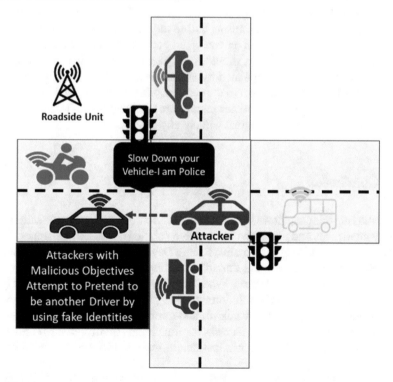

Fig. 4.7 Masquerading automotive attacks

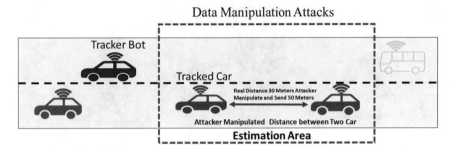

Data Manipulation Attacks

Fig. 4.8 Data manipulation attacks

4.6 Possible Solutions of Automotive Technology to Attack Types and Threats

To address the main types of cyber-attacks and threats we just looked at, there are five basic automotive cyber security solutions. First, **Authentication/Authorization** is used to establish trust with other entities. The vehicles need to know for a fact that the other entity is really who they claim to be. Second, **Encryption**, establishes trust in the overall process and mechanisms during the exchange of information. When the information is transmitted in an encrypted packet, an attacker cannot interpret the data. Third, **Data Integrity**, provides the necessary trust that the transmitted and received data is not manipulated by an attacker. Fourth, a **Digital Signature** helps to identify a trusted member in the network with negotiation and final contract rights. Fifth, **Non-Repudiation** which makes sure that other parties in networks have completed and followed the contract without any defects.

4.7 Conclusion

While studying this chapter we have seen the overview of advanced automotive cyber security challenges and also provides possible solutions of those challenges. It has provided the basic idea of Automotive Secure Development Lifecycle, Automotive Cyber Security: Challenges and Possible Solutions, Autonomous Cyber Security Risks, Automotive Cyber Attacks Types and Threats and the Automotive Cyber Security Layers and possible solutions. In this presentation, a concise review of ongoing outcomes has been introduced, and current cybersecurity and protection threats is portrayed remembering attacks for functional security, attacks at financial institutions, attacks on the network infrastructure, and considerations around security.

Acknowledgements This research was funded by Woosong University Academic Research in 2021.

References

1. NCCGroup, ASDL: automotive secure development lifecycle, report (2015)
2. European Automobile Manufacturers Association, ACEA principles of automotive cybersecurity, a report, Sept 2017
3. https://www.acea.be/uploads/publications/ACEA_Principles_of_Automobile_Cybersecurity.pdf
4. Altran report, Cybersecurity in automotive: how to stay ahead of cyber threats? Report (2018). https://www.altran.com/as-content/uploads/sites/5/2018/01/cybersecurity-in-automotive_position-paper.pdf
5. L.A. Vinh Hoa, A. Cavalli, Security attacks and solutions in vehicular ad-hoc networks: a survey. Int. J. Ad-Hoc Netw. Syst. (IJANS) **4**(2), 1–20 (2014)
6. M. Singh, S. Kim, Security analysis of intelligent vehicles: challenges and scope, in *2017 International SoC Design Conference (ISOCC)*, Seoul (2017), pp. 13–14. https://doi.org/10.1109/ISOCC.2017.8368805
7. M. Amin, Z. Tariq, Securing the car: how intrusive manufacturer-supplier approaches can reduce cybersecurity vulnerabilities. Technol. Innov. Manag. Rev. 21–25 (2015)
8. J. Liu, S. Zhang, W. Sun, Y. Shi, In-vehicle attacks and countermeasures: challenges and future directions. IEEE Netw. **31**(5), 50–58 (2017). https://doi.org/10.1109/MNET.2017.1600257
9. F. Sommer, J. Durrwand, R. Kriesten, Survey and classification of automotive security attacks. Information **10**(4), 148 (2019). https://doi.org/10.3390/info10040148

Chapter 5
Security Analysis of Intelligent Transport System

Madhusudan Singh ⓘ

Abstract Intelligent transport system cyber security deals and focusses on the problems related to cyber-security of transport system. As the transport system starts to get intelligent, more the vulnerable for cyber-attacks. Here we have basically dealt with cyber security for both intelligent vehicles and the infrastructure of intelligent transportation system. The overview of automotive requirement specification, Elicitation and validation with verification is highlighted in this review report. And more elaboration towards security requirement definition and method to improve the security requirement in advanced automotive technology.

Keywords Intelligent transportation system (ITS) · SCMS · ITS cyber security · Self-adaptive system

5.1 Introduction

The intelligent transportation system is the domain of the futuristic transportation system. ITS shall make use of advanced upcoming technologies such as wireless sensor networks. Distributed system architectures, human machine interface, and sensing and actuating control. The vehicles in an intelligent transportation system will be able to detect multiple types of vehicles on the for the world's roads [1]. ITS will improve safety for the passengers, optimized services transportation system and provide real time characteristics we will examine next. In figure shows the features of ITS.

Here in Fig. 5.1 advanced Intelligent Transportation System (ITS) is displayed equipped with sensors and used distributed system architecture. It shows that this technology is capable of well-defined human machine interface.

In Fig. 5.2 the equipped sensors in intelligent vehicle can detect multiple type of obstacle on real time basis and can provide real time traffic status.

M. Singh (✉)
School of Technology Studies, Endicott College of International Studies, Woosong University, Daejeon, Republic of Korea
e-mail: msingh@wsu.ac.kr

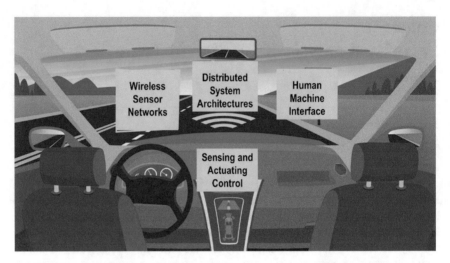

Fig. 5.1 Advanced ITS technologies

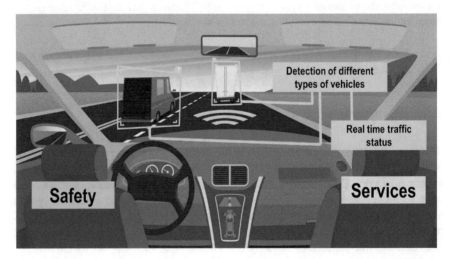

Fig. 5.2 The features of ITS

Before, street vehicles were autonomous and to a great extent mechanical framework. Notwithstanding, present day vehicles are progressively dependent on interior systems that connect sensors, actuators and control frameworks so as to accomplish more significant levels of usefulness than can be given by independent subsystems. Huge numbers of these capacities are wellbeing related, for example, antilock slowing down, cutting edge crisis slowing down, electronic strength control, and versatile voyage control. In the car condition, in this way, digital security assaults could have noteworthy wellbeing suggestions for vehicle inhabitants and other street clients, notwithstanding the protection and money related ramifications that are all the more usually connected with assaults on data frameworks. Such wellbeing issues could be accidental symptoms of some digital security assaults, yet it is possible that causing passing or injury could even be an essential assault objective for some expected assailants.

The rest of the chapter is organized as follows. Section 5.2 provides an overview of Intelligent Transportation System. Section 5.3 discusses the Cybersecurity challenges in ITS. Section 5.4 presents self-adaptive cyber security solutions with the USE CASE in ITS cyber security. Section 5.5 conclude an overview of current available ITS cyber security and future research directions.

5.2   Intelligent Transportation System: An Overview

An Intelligent Transportation System in order to be developed must have automation Information Technology Security in the form of a credential management system and knowledge sharing and finally it's use spectrum knowledge Transfer as presented in Fig. 5.3.

In below section, we have explained ITS order to be developed technology form.

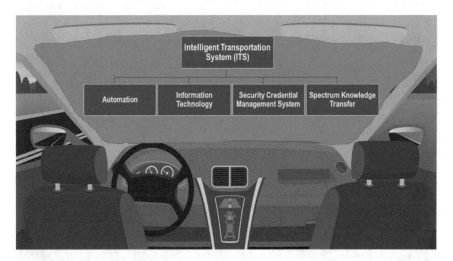

Fig. 5.3 Intelligent transportation system

5.3 Autonomous Vehicles on Intelligent Transportation System

The Intelligent transportation system provides the speed and distances of other vehicles on the road, real-time traffic and accident information, choice of the optimum route helpful CCTV images, even weather details and local events that might affect traffic. All this in time with automated features [2]. We can see example in Fig. 5.4.

The intelligent transport system uses many technologies such as wireless sensor networks, cloud computing, networking and communications etc. to collect and share real time data within the interconnected vehicles in ad-hoc network.

5.3.1 Information Technology

Information Technology helps to provide many services to the users, the ability to access information such as the vehicle operation route, current location, vehicle operation speeds at each location, estimated arrival time and the level of traffic congestion. Information technologies are associated with insightful transportation systems including remote correspondence, computational advancements, skimming vehicle information/drifting cell information, detecting advances, and video

Fig. 5.4 Automation in vehicle

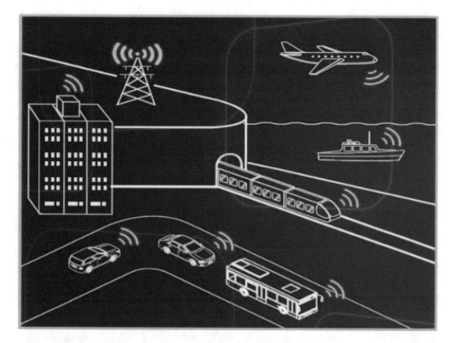

Fig. 5.5 Information technology in intelligent vehicle systems

vehicle discovery. Intelligent transport differs in the advances applied, from essential management system to more application systems, for example, crisis vehicle notice system, programmed street authorization, collision avoidance system, and some cooperative systems. Figure 5.5 has shown the IT communication method with different vehicles.

Here in Fig. 5.5 it has shown that how the intelligent vehicles exchanges data and information between them such as route, speed, location, etc.

5.3.2 Security Credential Management System

A security credential management system for vehicle, ITS correspondence is begun as joint endeavor by Crash Avoidance Metrics under a Cooperative Agreement with the United States Department of Transportation. This structure has formed altogether towards evidence of-idea from base scratch research and is driving possibility to help the foundation of an across the country model for framework plan for vehicle to intelligent transportation system security [3]. This system design issues a digital certificate for every vehicle on road who wish to participate on this project or infrastructure nodes for reliable and trustworthy communication among participant vehicles, which is the key to model for safety and mobility application that are based on vehicle to everything communication.

The Security Credential Management System (SCMS) is a proof-of-idea (POC) message security answer for vehicle-to-vehicle (V2V) and vehicle-to-system (V2I) correspondence. It utilizes a Public Key Infrastructure (PKI)-based methodology that utilizes incredibly imaginative procedure of encryption and declaration the executives to encourage confided in correspondence. Approved system members utilize advanced testaments gave by the SCMS POC to verify and approve the security and versatility messages that structure the establishment for associated vehicle innovations. To make sure about the insurance of vehicle owners, these supports contain no near and dear or apparatus perceiving information, yet fill in as structure affirmations with the objective that various customers in the system can trust in the wellspring of each message. The SCMS POC additionally plays a key capacity in ensuring the substance of each message by distinguishing and evacuating getting into mischief gadgets, while looking after protection.

5.3.3 Spectrum Knowledge Transfer in ITS

Finally, spectrum knowledge transfer stands for information delivery regarding roadside sensing which brings together tools and mechanisms that directly capture and convey data measurements from the road, obtaining valuable metrics such as speed, direction, flow of traffic and even which vehicles are traversing a given road segment, structured/static data, which refers to data sources that provide information of elements that have a direct impact on transportation, such as public transportation lines and timetables, or municipal bike rental services [4]. We can see spectrum knowledge transfer within ITS in Fig. 5.6.

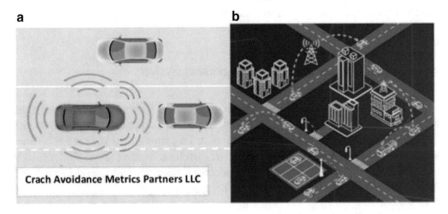

Fig. 5.6 a Security credential management system. **b** Spectrum knowledge information transfer within ITS

Fig. 5.7 Basic goals of intelligent system

ITS bandwidth that augments social government assistance could be either considerably more or significantly less than what has just been apportioned in light of the fact that ideal transmission capacity is delicate to unsure factors, for example, device infiltration, future data rates, and spectrum opportunity cost. That vulnerability is balanced if the ITS spectrum is shared. The data transfer capacity required to get given throughputs on the mutual range can be extensively not exactly the transmission capacity to get similar throughputs in different groups. The spectrum accessible for ITS ought to be kept up or expanded, yet quite a bit of ITS spectrum ought to be imparted to non-ITS devices.

5.3.4 Goals of Intelligent Transfer System

The whole use of ITS depends on information assortment, investigation and utilizing the consequences of the examination in the tasks, control and exploration ideas for traffic the executives where area assumes a significant job [5]. Here sensors, data processors, correspondence frameworks, side of the road messages, GPS refreshes and mechanized traffic prioritization signals assume a basic job in the use. As we can see in Fig. 5.7 has represented the basic goal of ITS.

The Goals of an Intelligent Transportation system can be listed in five categories. Those are representing in Fig. 5.8.

Fig. 5.8 Goals of an intelligent transportation system

In below, we have described each category with example.

- **Roadway Reporting**

The intelligent transportation system objectives are to exercise the high-volume traffic development progressively productive and improving street wellbeing. To accomplish these objectives, street administrators need to continually screen traffic and current street conditions. Street revealing is finished utilizing a broad exhibit of cameras and sensors that are deliberately positioned all over the street and which sends back information continuous to the control community. Models incorporate transport path cameras, speed cameras, side of the road climate stations, and vehicle detection system.

Each vehicle on the road shares monitored data with vehicles in ad-hoc network regarding traffic congestion, accidents, weather condition, etc. This is what called as roadway reporting as mentioned in Fig. 5.9.

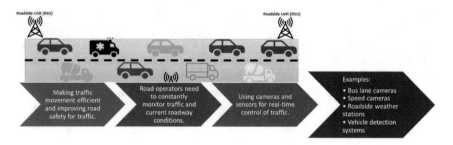

Fig. 5.9 Example roadway reporting each vehicle

Fig. 5.10 Vehicle sharing traffic flow controls in ITS

- **Traffic Flow Controls**

Like street detailing frameworks, traffic stream controls help in making high-volume traffic increasingly productive just as making streets more secure. The Road administrators screen traffic and street conditions continuously and utilize assembled information to oversee traffic stream utilizing different stream control instruments. The designed system includes traffic signal monitoring system, railroad path crossing management and various flow control mechanism. We have shown the traffic flow controls overview in Fig. 5.10.

Vehicles on the road constantly monitor space between vehicles and traffic congestion using rear as well as front cameras and the innovative technology of LIDAR. Every monitored data is shared within interconnected vehicles for traffic flow control.

- **Payment Application Systems**

Payment application and systems is a major ITS goal in order to manage and make the system smooth while organizing traffic movement and stretching towards efficient safety protocol. With these systems ITS operators use existing system of revenue collection to increase their revenue by significantly reducing costs. The most used methods are RFID tags, kiosk payment attribute, and e-ticket application. The next ITS goal we are going to discuss are management application systems. Figure 5.11 has given an overview about toll payment system.

In the highways and toll tax center area vehicles need to pay tax or fines if they break any systematic rule. These issues are maintained in ITS (as seen in Fig. 5.11)

Fig. 5.11 Example of payment application system and recognized vehicle based in ITS camera

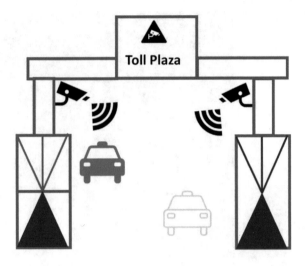

by using camera and sensors. Basically RFID tags are placed on the vehicle and when they move across the highway sensor and camera reads the RFID and charges the amount through payment application system.

- **Management Application Systems**

Management Applications and Systems for ITS is quality defined as a complex environment comprising of hundreds and hundreds of connected structures with a wide variety of functions, running cooperatively or in tandem. All system wants to operate within defined tolerance limits for visitors to glide smoothly; otherwise, there'll be traffic jams, delays, and accidents. ITS nerve facilities host, monitor, and perform the management systems controlling ITS. Models: streetlight controls, catastrophe the executives, information and information stockpiling the executives, crisis vehicle the executives, and traffic and clog the executives. The management application system has presented in Fig. 5.12.

Centralized management system controls every subsidiary management system distributed across the ITS infrastructure in which thousands of connected system exist and have wide range of functions.

- **Communication Application System**

Communications application and systems provide the Information alternate that's the center of the ITS ecosystem. Data is used for making site visitors waft efficient, enhancing road safety, growing revenue, and lowering the ecological and environmental impact. Data is additionally devoured by the clients of ITS administrations to improve their travel alternatives and encounters. Instances of such uses are brilliant applications, internet-based life, sites, street obstruction and mishap alerts.

In Fig. 5.13 it has shown how every IoT devices is connected to exchange and share data within the ad-hoc network for convenient and smooth ITS system. It shares

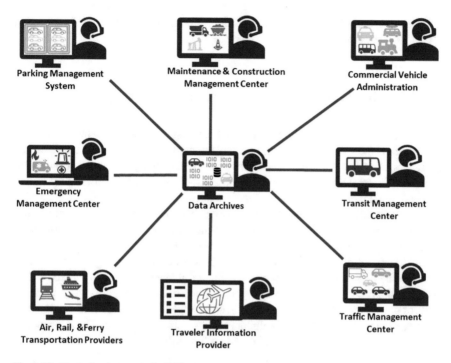

Fig. 5.12 Centralized controls for ITS management system

Fig. 5.13 Example of communication application systems

data between the devices and vehicles to alert regarding each and every factor either it is weather or temperature or any thing else.

5.4 Benefits of Intelligent Transportation System

When showcasing the benefits of the intelligent transportation system the first usually mentioned is the vast Improvement in safety which is measured by the reduction and mitigation of accidents. Another is the increased connectivity of vehicles which helps to expand the capacity of roadway infrastructure, enhances traffic flows and provides more personal mobility options for disabled and aging populations [6]. Last, another important benefit is the reduction in energy consumption thanks to low emissions that "occur as a result of aerodynamic drafting" and the "overall improvement of traffic flow dynamics". It has represented in Fig. 5.14.

Intelligent Transportation Systems (ITS) consolidate a wide range of sorts of data and interchanges innovation to make a system of frameworks that help oversee traffic, secure streets and that's only the tip of the iceberg. As an ever-increasing number of parts of our transportation arrangement become organized, ITS will change the way drivers, organizations, and governments manage street transport. These propelled systems can help improve transportation in a few different ways.

- **Better Traffic Safety**: High speed, unexpected climate conditions and substantial vehicle congestion would all be able to prompt mishaps and the death toll; intelligent transportation frameworks help with these. Constant climate checking frameworks gather data on visibility, wind speed, rainfall, street conditions and the sky is the limit from there, permitting traffic controllers authorized data on driving conditions. In completely organized frameworks, this data would then be able to be utilized to refresh notice signs and even speed restrains when the need emerges, keeping drivers caution to the conditions around them. Crisis vehicles can react

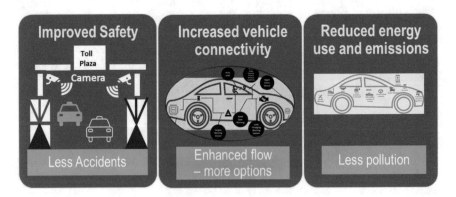

Fig. 5.14 Benefits of intelligent transportation systems

rapidly to mishaps as constant traffic checking cautions them. ITS traffic control occupies traffic away from occupied or perilous regions, forestalling congested driving conditions yet in addition lessening the danger of impacts.

- **Decreasing Infrastructure Destruction**: Overwhelming traffic congestion can put a great deal of system overload out and about system, especially when they're over-burden. Checkpoints and other more established types of weight control diminish the danger of over-loading but at the expense of wasted time postponed traffic. Weigh in motion measure the sort, size and weight of vehicles as they move, imparting the gathered information back to a central server.
- **Traffic Control**: The intelligent transportation system, be that as it may permit traffic lights to react to changing examples themselves. Versatile traffic light frameworks make savvy crossing points that control traffic in light of the examples they see among the vehicles utilizing them. They can likewise organize explicit types of traffic, for example, crisis vehicles or open travel.
- **Parking Management**: Drivers acknowledge they will normally be alluded to for illegal or expanded halting. These customized structures help improve traffic stream by extending driver consistence and smooth turnover of parking spaces.
- **Gathering Traffic Data**: The intelligent transportation system can quantify the number and kind of vehicles utilizing a specific street or be visiting a specific piece of a city, just as pinnacle traffic times, venture length, and other information. This data assists specialists with assigning their assets in the most proficient conceivable way.

5.5 Intelligent Transportation System Security Challenges

Intelligent transportation system has also a lot of security challenges. It can be categories three section: Physical Attacks, Wireless Attacks, Network Attacks. It is truly conceivable that a solitary attack can include every one of the three attack classifications simultaneously. For instance, in an attack against a traffic stream controller gadget, for example, a Dynamic Message Sign (DMS), the assailants can send the gadget off base/inappropriate orders through a remote connection, by truly interfacing with the gadget, as well as over the system using a sabotaged controller application. Nature besides, the value of DMS makes this attack vector multimodal [7]. We can get clearer image of each of the 3 attack in underneath.

Physical Attacks: ITS foundation truly uncovered on streets and side of the road, making them available to any individual who approaches them. The ITS foundation sits truly uncovered on streets and side of the road, making them available to any individual who approaches it. Attack vectors that may mishandle this ITS simplicity ease-of-physical access nature may sits truly uncovered on streets and side of the road, causing it to include physically interfacing with revealed ports e.g., USB.PS2.serial, etc. Sniffing system traffic between a contraption and the backend, Scanning the ensured about/shut framework to discover it's geology, Man-in-the-Middle attacks utilizing any revealed wires/connections to get data, truly meddling with a device to

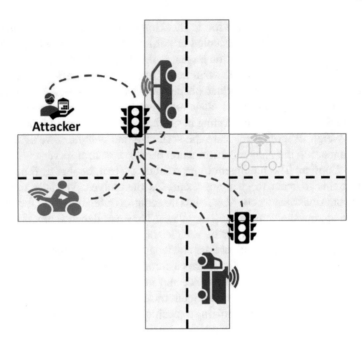

Fig. 5.15 Physical attacks

take/deal data, modifying a device, etc., connecting a removable stockpiling gadget stacked with malware to introduce and Turning an ITS device as an accepted area point into the corporate system [8].

In Fig. 5.15 we can see how any individual can attack any element of inrastructure of ITS. Like here a person is trying to attack on traffic light. It might break the coordination of traffic system and dta regarding that particular point traffic system might not reach to the destination where expected.

Wireless Attacks: The complete vehicular networks (V2V, V2I, and V2P) are the communication dedicated for vehicle short range communication. These are the backbone of future ITS operations. As the network will start to flex the vulnerability to get attcked increases, and the ITS infrastructure poses a major security challenge for ITS operators. There are atleast five attacks vectors which are major threat in wireless transmission. Spoofing is very probable which exhibits due to the model design of vehicle to vehicle and vehicle to infrastructure. Messages broadcasted to traffic may be compromised by exploiting vulnerabilities in the software used, the actual hardware devices, the protocols, the operating systems, etc. man in middle attacks the wireless transmmisiion to intercept and/modify data. As seen in the graphic our hacker establishes a new parallel connection to the oroginal connection the last wireless attack we will examine involves attacking the car's Wi-Fi using malicious app installed in phone connected to it via bluetooth or wifi for portable usability. Figure 5.16 can see the wireless attacks in the following sequences.

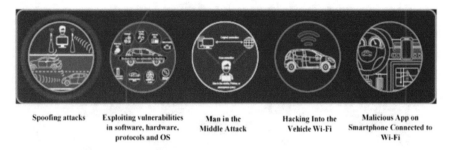

Spoofing attacks   Exploiting vulnerabilities   Man in the   Hacking Into the   Malicious App on
                    in software, hardware,   Middle Attack   Vehicle Wi-Fi   Smartphone Connected to
                    protocols and OS   Wi-Fi

Fig. 5.16 Multiple possible wireless attacks

Any kind of internet theft excluding physical can take place as shown in Fig. 5.16. It might be spoofing to harm n another basis or data theft by any middle man or just to break the communication.

Network Attacks: This is an area wherein most of the cyberattacks against structures can begin once again the system itself. Web uncovered the ITS systems, discoverable by methods for IoT web search instruments, for instance, Shodan, unequivocally assaults. The standard framework based assault vectors fuse, Identifying and mistreating contraption misconfigurations, far away machine exposure and abuse, c.G., disclosure using Shodan, Installing malware/adware on frameworks, and present-day domain supported focused assaults or advanced persistent threats (APTs). Social engineering attacks. E.G. Spear-phishing. Now that we mentioned network, physical and wi-fi attacks, we can recognition on unique attacks in opposition to vehicular advert hoc networks. Inside the near destiny, will become the number one roadway users.

Connected cars of today and the autonomous cars, one of the important thing technologies related motors will use, are vehicular advert hoc networks. These systems are made out of smart vehicles and Roadside Units which talk through questionable remote media Vehicular specially appointed systems are made out of smart vehicles and Roadside Units which talk through inconsistent remote media. These systems are powerless against attacks that can endanger street wellbeing, particularly while vehicles rely upon their measurements for settling on significant riding decisions. Let us see some genuine worldwide instances of ITS attacks. The hacked transportable motorway message board inside the pictures above is a case of a real worldwide ITS ambush. These assaults are occurring today, the interconnected ITS environment that we characterized already isn't however totally executed. We are as a base 10 years from completely interconnecting each vehicle and every street ITS system.

We can most effective imagine what will happen inside the destiny while increasingly more connected ITS systems can be coming online Such attacks will be proportionally growing. ITS will really need to contend with digital attacks as entire incorporation and interconnection is gradually performed. This burdens the significance of making sure about ITS from the beginning. All things considered. ITS will not just

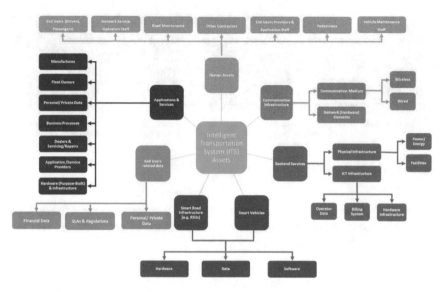

Fig. 5.17 Intelligent transportation system assets

affect financial benefits however on open wellbeing as appropriately. The components of ITS assets have shown in Fig. 5.17.

The ITS infrastructure is broad and widely distributed and it tends to be interconnected with every single hardware device within the IoT platform.

The fuse of knowledge is turning into an inexorably significant piece of the transportation framework. Attacks on the transportation foundation have been constrained, however as more vehicles become associated, the danger for digital attacks increments and thus the need to make sure about Intelligent Transportation Systems (ITS) for singular vehicles and open transportation. The security of these frameworks is significant for sheltered and proficient transportation. Strategies and calculations should critically address the vulnerabilities of system and conquer the outside dangers to the system, and to lessen the dangers of attacks that can be looked by ITS.

5.6 Self-adaptive System

The owned vehicle environment is dynamic in nature and very vulnerable to attacks, static security cannot change the security mechanism and make the possible right decision at runtime. To solve this problem, dynamic security or reconciling security is required. Here we will focus towards self-adaptive system requirements for connected vehicles to understand the security measures and issues in security in self adaptive system. Let's look at the current upcoming advance intelligent technologies for vehicles such as in the case like driverless vehicles.

A self-adaptive system satisfies an anticipated model prerequisite by detecting the environment, dissecting it, and making the best choice at the current case situation as indicated by the decided circumstance around then situation. Model can prepare for its own decision for the best case in real time which requires a strong understanding of all the trade-offs among the different requirements. Self-adaptive system is well trained before the launch and has the complete consciousness of its architecture and model and along these lines it reconfigures self-ruling at runtime to initiate just the necessary modules for a specific domain state.

The self-adaptive system is connected to the cloud ITS platform. It relies on sensors that are electronic components that collect the data and a special software-device combination that can react in response to stimuli and execute decisions. These software enabled devices are called effectors. We have thousands of sensors in vehicles and the vehicular environment in general and multiple of effectors that can act and make decisions. The adaptation manager monitors, analyzes plans and executes the decisions.

Intelligent Transportation System (ITS) is the developing transportation which contains the trend setting innovations, for example, remote correspondences, video vehicle discovery, appropriated system design, human machine interface, detecting and inciting, to improve security of the travelers, traffic clog, fuel utilization and streamline different administrations of the transportation framework, for example, constant traffic circumstance, electronic cost assortment, programmed street requirement and hot paths. The current ITS frameworks needs self-versatile capacity to take unconstrained choices. Nature of the ITS framework is extremely intricate so ideal choices can't be taken however a sufficient choice can be taken which satisfies the requirements of the system.

All these decisions are derived from the knowledge database. The model is known from the first letters of each action MAPE-K [9], which as explained mean: Monitor, Analysis. Plan and Execute. The letter K stands for the knowledge from knowledge database. The adaptation is implemented through continuous feedback loop as represented in the diagram here with the blue arrows. The persistent checking movement recognizes possible changes, and the system will continue playing out each of the four exercises in a ceaseless circle. The components of the adoption manager have shown in Fig. 5.18

- **Monitoring (M)**: Monitoring the self and the surrounding. In this piece of the adaptation manager, the system utilizes information gathered by the sensors to assemble crude data about the vehicles and the general vehicular condition.
- **Analyzing (A)**: Analyzing the monitored data. In this part of the adaptation manager the system analyzes the raw data for the purpose to gain understanding and the semantics, raw data is structured and becomes information.
- **Plan (P)**: Formulating a decision plan. In this part of the adaptation manager the system uses the information constructed in the previous phase to create a reasonable plan of action based on the pre-determined actions in the knowledge database. It automatically selects the optimum suitable acts or series of actions.

Fig. 5.18 The elements of the adoption manager

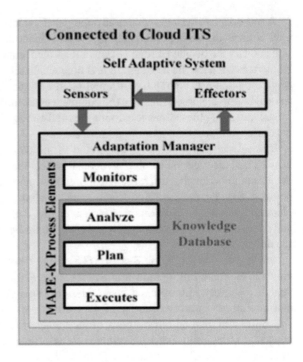

- **Execute (E)**: Executing the decision plan. In this part of the adaptation manager the system executes the plan via the effectors. The plan is dynamic and keeps adapting based on environment related changes.
- **Knowledge (K)**: Knowledge database of the system. In this part of the adaptation manager every potential planning activity is pre-stored. The system when in the analysis and planning phases accessed the knowledge database to perform reasoning tasks that it believes to be appropriate.

In Fig. 5.12 we can see MAPE constitutes the knowledge element. Initially the data is monitored using sensors and actuators and then analyzed. After complete analysis using either machine learning algorithm or other then the plan is set up and the execution is completed through effectors.

The system main requirement is safety therefore, if someone breaks any traffic rule and creates nuisance on the road then the system suspects the person breaking rule or creating any nuisance on the road then the system presumes the individual defying norm and alarms different drivers in the scope of surprising occasion.

5.7 Adaptive Cyber Security for Autonomous Vehicles: USE CASE

Security systems for intelligent vehicles is an emerging research topic and therefore adaptive system security requirements must be identified. An autonomous vehicle (AV) is a vehicle that works and performs undertakings under its own capacity. A few highlights of the self-ruling vehicles are detecting the earth, gathering data, and overseeing correspondence with different vehicles. Numerous self-ruling vehicles being developed utilize a mix of cameras, sensors, GPS, radar, LiDAR, and onboard PCs. These advancements cooperate to plan the vehicle's position and its nearness to everything around it. In light of their dependence on such advancements, which are effectively open to altering, independent vehicles are vulnerable to digital attacks if an attacker can find a shortcoming in a particular sort of vehicle or in an organization's electronic system. This absence of data security can prompt criminal and terrorist acts that in the eventually cost lives. The vehicle system should be able to first sense and analyze the surrounding environment embodied many intelligent vehicles and obstacles and particularly pedestrians every second for decisive execution. Adaptive system security goals are to monitor analyze changes in the system, and ultimately provide countermeasures at a satisfactory level of protection.

In this scenario, we will explain how the self-adaptive the MAPE-K method we described previously can be used for automotive cyber security purposes. This model can provide end to end encrypted model of communication on real time basis. This is because it can provide self-adaptive solutions for different security challenges like at communication layer eavesdropping, spoofing, man-in-the-middle, and sybil attacks are most targeted kind of attacks which releases our privacy and information. Security challenges like risk recognizable proof, message confirmation, trustworthiness, non-revocation, and attack against privacy. The case scenario of when two intelligent vehicle 'A' and 'B' exchanges information, i.e. when A receive message from B. Vehicle 'A' can directly establish communication with vehicle 'B' and start communication using adaptive MAPE-K security model. This model generates effective result after execution regarding security based on certain criteria and algorithm.

We can see in Fig. 5.19 how IV(A) and IV(B) interact and share data through MAPE-K structure.

The adaptation manager is able to use the monitoring part to check the sender's validity. So in this case vehicle 'A' can check whether vehicle 'B' is a legitimate vehicle enlisted with the authentication community and its obligation and previous history to inspect the risk. Also, it can legitimately check vehicle 'B's testament on its own kept up a database that containing the database of all recently imparted clever vehicles on the cloud or it can, or it can't confirm legitimacy with the assistance of an endorsement, for example, a computerized signature. As presented in Fig. 5.20.

Both intelligent vehicles 'A' and 'B' can use th analysis part of the adaptation manager to analyze the authentication of the messages exchanged. Here the core value of this rule/algorithm is to anlayze regardless of whether the message got from

Fig. 5.19 Adaptive cyber
security for autonomous
vehicles: a USE CASE

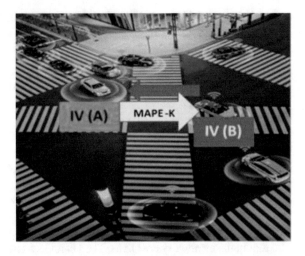

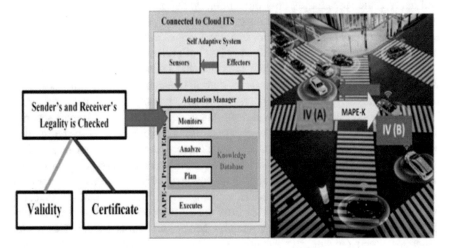

Fig. 5.20 Monitor functionality

the originator is real and furthermore at what degree it is guaranteed is that the
message isn't blocked while in transit.

After monitoring stat system analyzes the authentication message received from
another end vehicle. Here IV(A) authenticates IV(B) message (Fig. 5.21 presented).

The receiver intelligent vehicle 'A' depends on the secured and end to end encrpted
communication model with sender intelligent vehicle 'B' by using cryptographic
protocol that secures the communication to maintain the pricavy of the receiver.

After the message authentication a plan is sketched and messages are sent
according to plan as shown in Fig. 5.22.

The messages for both vehicles are sent according to plan and cannot be repudiated
which that the either side of communication means i.e. Vehicle 'A' or 'B' sender

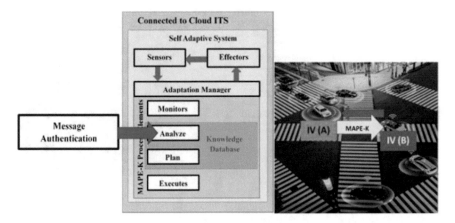

Fig. 5.21 Analysis process

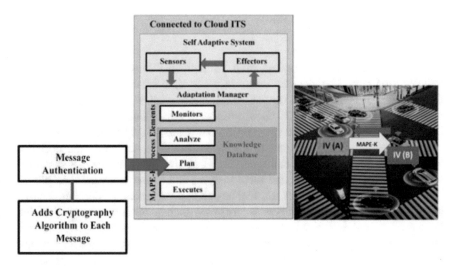

Fig. 5.22 Plan process

or recipient can't deny any message or data trade between them. All layer of the adaptive MAPE-K security model will be exercised based on the algorithm defined on the model and it's data base.

Knowledge data base stores the data after analyzing the monitored data. The planning activity uses the knowledge of the environment. In the knowledge database predefined rules are stored as shown in Fig. 5.23.

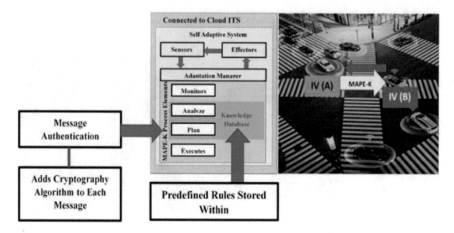

Fig. 5.23 Execute process access the information from knowledge database

5.8 Summary

This article presents an overview of the Intelligent Transportation System, it's features, goals and benefits showcased in a real time example. We also explored the potential of cyber-attacks on ITS technologies such as wireless, network and physical attacks. We also learned about attacks against vehicular networks and examined some real-world examples of cyber-attacks against ITS systems. We showcased self-adaptive systems and engineering requirements for Intelligent Transportation Systems. We learned MAPE-K to apply self-adaptive model in the ITS system in which first they can monitor and then analyze, and as per analyzed data plan is sketched and further executed. Last, we showed how we can use reconciling security to solve issues in the adaptive cyber security environment for Autonomous Vehicles. We have seen how Self-compromise in necessity designing is gained by five structure squares, for example, Monitor, Analyze, Plan, Execute and Knowledge for example [MAPE-K] circle design, which gives the accommodating abilities to our proposed accommodating prerequisites for digital security in IV.

Acknowledgements This research was funded by Woosong University Academic Research in 2021.

References

1. H. Sedjelmaci, M. Hadji, N. Ansari, Cyber security game for intelligent transportation systems. IEEE Netw. **33**(4), 216–222 (2019). https://doi.org/10.1109/MNET.2018.1800279
2. G.-U. Rehman, A. Ghani, S. Muhammad, M. Singh, D. Singh, Selfishness in vehicular delay-tolerant networks: a review. Sensor **20**, 3000 (2020)

3. R. da Rosa Righi, et al., Reducing cost and time-to-market on supporting driver assistance systems to avoid rear-end collisions in vehicles traffic, in *2019 IEEE International Conference on Computational Science and Engineering (CSE) and IEEE International Conference on Embedded and Ubiquitous Computing (EUC)*, New York (2019), pp. 367–372. https://doi.org/10.1109/CSE/EUC.2019.00076

4. A.R. Ruddle, D.D. Ward, Cyber security risk analysis for intelligent transport systems and in-vehicle networks, in *Intelligent Transportation Systems* (Wiley, 2015), pp. 83–106. https://doi.org/10.1002/9781118894774.ch5

5. M. Singh, S. Kim, Reconcile security requirements for intelligent vehicles, in *2017 17th International Conference on Control, Automation and System (ICCAS 2017)*, Ramada Plaza, Jeju, South Korea

6. S. Kwag, S. Lee, A survey of V2X communication technologies and project. J. Korea Soc. Automot. Eng. **33**(5), 24–31, May (2011)

7. M. Singh, S. Kim, Security analysis of intelligent vehicle: Challenges & scope, in *14th International SoC Design Conference (ISOCC 2017)*, Grand Hilton Hotel, Seoul, South Korea, Nov. 5–8, (2017)

8. J.R. Reagan, M. Singh, Automotive evolution. In *Management 4.0. Blockchain Technologies*. Springer, Singapore (2020). https://doi.org/10.1007/978-981-15-6751-3_2

9. M. Singh, Tri-Blockchain based intelligent vehicular networks, in *IEEE INFOCOM 2020 - IEEE Conference on Computer Communications Workshops (INFOCOM WKSHPS)*, Toronto, ON, Canada, pp. 860–864 (2020). https://doi.org/10.1109/INFOCOMWKSHPS50562.2020.9162692

Chapter 6
Cryptographic Techniques for Automotive Technology

Md. Iftekhar Salam and Madhusudan Singh⊙

Abstract Modern vehicles now can be connected over vehicular network that gives the opportunity for building an intelligent transportation system. These networks provide many attractive features to provide comfort and safety. However, this also raises a new set of security concerns. These security features must not be limited to protecting confidential information, but also needs to address safety critical systems such as brake, accelerator or steering etc. this article has presented the cryptography mechanism overview for automotive cybersecurity.

Keywords Automotive cryptography · Stream cipher · Symmetric encryption

6.1 Automotive Cybersecurity

Not long ago, in 2015 two automotive security researchers Charlie Miller and Chris Valasek demonstrated a remote attack on the Jeep Cherokee by taking control of the Jeep's brakes and accelerator. It turns out that the researchers exploited a vulnerability in the Jeep's infotainment system over a cellular network. Imagine an attacker hacking into the electronic component of the vehicle to get access to the control of the braking system; the result would be devastating. Addressing the security of vehicular network is a challenging issue [1]. In Fig. 6.1, we have shown need of security in vehicle such as vehicle location, communication channel (Wi-Fi, cellular networks etc.) embedded insurance aggregator, vehicle valuation, driving behavior of vehicle, vehicle tracking, virtual breakdown on-demand, deals accident information, and virtual mechanic.

Cryptographic techniques play an important role to provide different security features in a vehicular network. Security of such network depends on various things,

Md. I. Salam
Department of Information and Communication Technology, Xiamen University, Xiamen, Malaysia

M. Singh (✉)
School of Technology Studies, Endicott College of International Studies, Woosong University, Daejeon, Republic of Korea
e-mail: msingh@wsu.ac.kr

© The Author(s), under exclusive license to Springer Nature Singapore Pte Ltd. 2021   85
M. Singh, *Information Security of Intelligent Vehicles Communication*,
Studies in Computational Intelligence 978,
https://doi.org/10.1007/978-981-16-2217-5_6

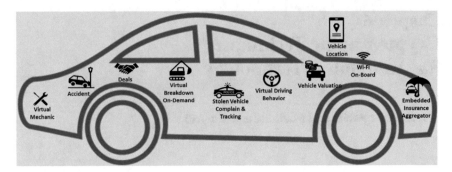

Fig. 6.1 Need of security in vehicle issues

but cryptography plays an important role as the basic building block for providing security in vehicular networks. Cryptographic techniques can be used to support the security goals such as authentication, confidentiality, integrity assurance etc., in V2X communications, where a vehicle is communicating with other vehicles (V2V) or roadside infrastructure (V2I) [1]. Cryptography bolsters the validation that permits these applications to confide in each other, which clearly is key since they include human lives in huge amounts of metal moving at high speeds. Crypto algorithms are additionally an incredible method to encrypt V2X interchanges. A model is guaranteeing that a vehicle's continuous area data has not been controlled. Without cryptography, it's conceivable that a programmer could send counterfeit message that could cause, for example, mishaps by activating programmed slowing down at high speeds.

In general, there are mainly two types of cryptographic primitives: symmetric primitives and asymmetric primitives. Symmetric cryptosystems are commonly used for providing security services of confidentiality and integrity assurance. Asymmetric cryptosystems are more commonly used for providing user authentication and key distribution.

- *Confidentiality* can be achieved using a secure symmetric key cipher that provides encryption and decryption functionalities. These symmetric ciphers are mainly categorized into two types: block ciphers and stream ciphers. Over the years there have been several developments of secure block ciphers and stream ciphers. Currently, the most common block cipher algorithm is Rijndael, which was selected as the Advanced Encryption Standard (AES). Block ciphers have different modes of operation to provide different services. Examples of block cipher mode of operations providing confidentiality include AES Cipher Block Chaining (AES-CBC) mode [3], AES Counter (AES-CTR) mode. Similarly, stream ciphers can also be used to provide confidentiality. Examples of stream cipher-based confidentiality algorithms include Salsa20, Trivium.
- *Integrity* assurance can be achieved using a message authentication code (MAC) algorithm or using a cryptographic hash function. There have been several developments of secure algorithms, e.g., Hash-based Message Authentication Code

(HMAC), Poly1305, to provide integrity assurance of the transmitted data. Modes of block cipher, e.g., Cipher Block Chaining MAC (CBC-MAC), can also be used to provide integrity assurance. Similarly, there are some stream cipher-based constructions, e.g., ZUC, which provide integrity assurance.

Generally, stream ciphers are faster than their block ciphers counterpart, and are most suitable for real time application. On the other hand, block ciphers are more suitable for processing bulk amount of data such disk encryption. In regard to the security goal of confidentiality stream cipher may seem to be a more suitable choice for automotive cybersecurity. This is due to requirement of several facts such as real time processing, low computational overhead, and faster processing.

The rest of the chapter is organized as follows. Section 6.2 presents an overview of symmetric cryptosystem-based stream ciphers. Section 6.3 discusses current research trend on symmetric ciphers. Section 6.4 discussed standard IEEE 1609.2. Section 6.5 presents the future of automotive cryptography. Section 6.6 conclude an overview of current automotive cryptography and future research directions.

6.2   Symmetric Cryptosystem Based on Stream Ciphers

Stream ciphers are broadly utilized in the cryptographic algorithms for giving confidentiality to negotiate the transmission of data between two clients. A stream cipher ordinarily partitions the message into progressive characters and works on each character independently to encrypt/decrypt the message. In view of the size of the character, a stream cipher can be either bit based or word based depending on the length of the character and its size. In the bit-based stream cipher, the cipher works on each bit independently. In the word-based stream cipher, each character comprises of a gathering of bits called a word and the cipher works on these words to encrypt/decrypt a message. Before going to the details of a stream cipher [2]. Figure 6.2 has shown the process of symmetric cryptosystem based on stream ciphers.

6.2.1   Notation and Terminology

- **Keystream generator** A component that generates pseudo-random characters.
- **Secret key, 'K'** An input component which entered in the keystream generator, which is only known to the sender and receiver.
- **Initialization vector (IV), 'V'** An input to the keystream generator, which is usually publicly available information. The initialization vector usually varies from message to message.
- **Keystream, 'Z'** Stream of output bits/words from the keystream generator.
- **Plaintext, 'P'** Stream of plaintext message bits/words before encryption.
- **Ciphertext, 'C'** Stream of the ciphertext message bits/words after encryption.

Fig. 6.2 Process of
symmetric cryptosystem

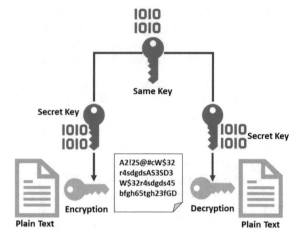

- *Associated Data, 'D'* Stream of the associated data bits/words. This part of a message does not require confidentiality; but requires integrity assurance.
- *Message, 'M'* Message can be either plaintext or ciphertext or associated data.
- *Tag, 'τ'* A specific length sequence generated by the tag generation algorithm. The tag is computed on the message value and is used to determine whether the message has been modified during transmission.
- *Encryption algorithm* A process that converts the plaintext into ciphertext using a secret key.
- *Decryption algorithm* A process that converts the ciphertext into plaintext using a secret key.
- *Internal state* Memory locations where information is stored.
- *Internal state size* Amount of information that the internal state of the cipher can hold.
- *Initialization* The initialization phase loads and disuses the key and initialization vector are stacked into the internal state of the keystream generator. The state obtained after the initialization procedure is called the initial state.
- *State updates function* the process to update the contents of the internal state.
- *Output functions* the process to compute the keystream bits using the contents of the internal state.

6.2.2 Operations to Provide Confidentiality

Confidentiality guarantees that the message isn't unveiled to an unauthorized entity. This can be accomplished by utilizing an encryption/decryption algorithm. Here, we quickly characterize the encryption/decryption method for accomplishing confidentiality utilizing a stream cipher. Figure 6.1 shows the general development of a stream cipher. Initially, the secret key, K, and initialization vector, V are stacked into

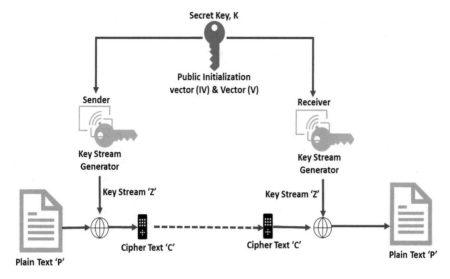

Fig. 6.3 Confidentiality using stream cipher

the internal state of the keystream generator as a major aspect of some initialization phase [3]. Following this, the keystream generator is operated for a predetermined number of iterations without creating any keystream bits. After the initialization phase, the internal state of the keystream generator comprises the initial state and is prepared to produce the keystream bits.

For keystream generation, the internal state of the keystream generator is refreshed utilizing a state update function and a keystream bit/word, 'Z', is computed at every iteration utilizing the output function of the cipher. The keystream generator is utilized to generate a keystream sequence by repeating this procedure. In conclusion, as appeared in Fig. 6.3, the encryption algorithm utilizes a combining function to consolidate the keystream Z with the plaintext P and outputs the ciphertext 'C'. Regularly, the keystream is combined with the message utilizing bitwise XOR function. Stream ciphers using the XOR function as the combining function are called binary additive stream ciphers. Upon encryption of the plaintext, the ciphertext is transmitted through the insecure channel. At the receiver end, the keystream is generated in a similar fashion and then the decryption algorithm combines the ciphertext C with the keystream Z to retrieve the plaintext.

6.2.3 Operations to Provide Integrity Assurance

Integrity assurance of a message gives the beneficiary an assurance that the information has not been altered during transmission. Data integrity assurance can be accomplished by producing a Message Authentication Code (MAC) tag. Here, we quickly portray a method for accomplishing integrity assurance utilizing a stream figure. In a stream cipher-based MAC tag generation scheme as appeared in Fig. 6.4, the input message M is collected into the internal state of the cipher subsequent to performing out the initialization phase. Following this, the cipher is iterated for a predefined number of steps in the finalization phase without creating any output bits. Toward the end of the finalization phase, the tag generation function takes input from a portion of the internal state bits and output the MAC tag τ [4].

The MAC tag is affixed and afterward transmitted with the message. Upon receipt of the message, the receiver processes the MAC tag τ' for the got message M' and compares it with the received MAC tag. On the off chance that the got MAC tag coordinates the registered MAC tag at the recipient ($\tau = \tau'$), at that point the receiver expects that the message has not been adjusted during transmission. In the event that the got MAC tag doesn't coordinate the MAC tag computed by the recipient, at that point the receiver expects that the message has been adjusted during transmission through the channel and therefore disregard the received message.

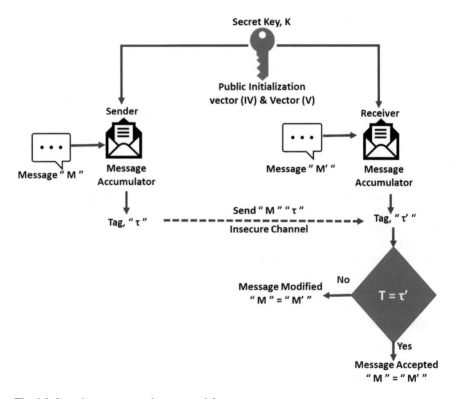

Fig. 6.4 Integrity assurance using stream cipher

6.3 Current Research Trend on Symmetric Ciphers

There are several cases where both confidentiality and integrity assurance are needed. The idea of providing both data confidentiality and integrity assurance is not new. A scheme that provides both security goals is called an authenticated encryption (AE) scheme. Individual algorithms can be combined to construct an authenticated encryption scheme. A generic composition to shape such authenticated encryption conspire was portrayed by Bellare and Namprempere, and Katz and Yung. These generic compositions are two-pass schemes, in this manner require two passes over the data: one giving secrecy and the other, integrity confirmation. These AE schemes also require two different keys, one for each of their component mechanisms. The computational cost of the two-pass scheme is about twice of the single pass scheme.

Authenticated encryption is one of the hot research topics these days. More importantly, we require authenticated encryption for lightweight devices such as sensor network, IoT devices. The lightweight cryptographic primitives will preferably be suitable for the automotive cybersecurity as well.

In view of this, starting from 2013 the "Lightweight Cryptography Project" was initiated in view to to design lightweight cryptography by National Institute of Standards and Technology (NIST). The aim of the lightweight cryptography project is to evaluate and standardize cryptographic algorithms which are suitable for resource constrained environment. This standardization process aims to select a portfolio of secure and efficient lightweight cryptographic algorithms.

6.3.1 Asymmetric Cryptosystem

Asymmetric cryptography [5], otherwise called public key cryptography is for the most part utilized for key conveyance and to give the security administrations of non-repudiation and client verification in vehicular networks. Asymmetric cryptosystems can likewise be utilized to give the security objective of confidentiality; nonetheless, slower and resource-hungry similar to the symmetric cryptosystem as shown in Fig. 6.5.

In this scheme, a couple of keys are utilized: a public key and a private key. In spite of the fact that the keys are different, the keys are numerically related. One key is called public key since it is known to everybody, while the other one is called private key just known to the proprietor. Consequently, Asymmetric Key Cryptography is otherwise called Public Key Cryptography.

- **Asymmetric cryptography for providing confidentiality**
  In this plan, each client must have a couple of various keys: a private key, and a public key. One implied for encryption of the information while the other implied for its decryption. The public key is placed into a public storehouse while the private key is put away as a mystery. Even though public and private keys are mathematically related, it is computationally unrealistic to get one from another.

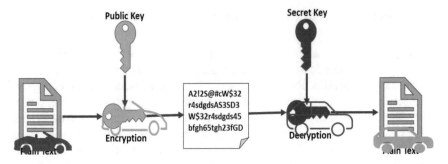

Fig. 6.5 Asymmetric cryptosystem

At the point when a client needs to send information to another client, at that point, they get the public key of another client from the store, encrypts the information, and afterward transmits it. Another client at that point utilizes their private key for decryption.

One inquiry you may pose, that, how these keys are identified with one another but then, it is difficult to get one from another. The appropriate response lies in mathematical ideas and concept. Numerically, it is conceivable to plan a cryptosystem whose keys have this property. Symmetric cryptography was fine government and military organizations, however with the spread of secure computers networks, an alternate sort cryptography strategy was required to address the issue as shown in Fig. 6.6.

- **Asymmetric cryptography for user authentication**
  A Digital Signature can be produced utilizing asymmetric cryptography to give client verification and non-repudiation in vehicular systems. This is a method that connects an element with the digital data which is supposed to transmit and receive. This affiliation is freely verifiable by the recipient just as any outsider or third-party interaction. Digital Signature is a cryptographic worth which can be determined from the data and some secret (key shared or belonged) known to the

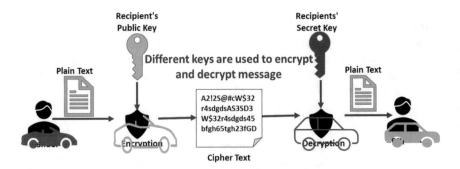

Fig. 6.6 Encryption and decryption of asymmetric key encryption

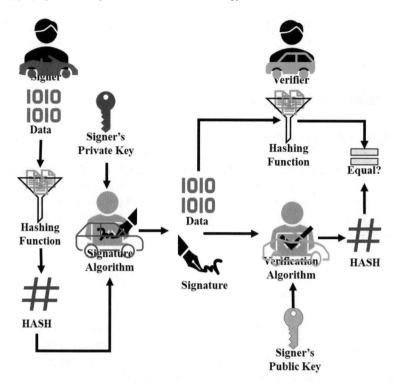

Fig. 6.7 Model of digital signature

signer. The digital signature scheme depends on public-key cryptography. The model of Digital Signature scheme is illustrated in the following Fig. 6.7.

Each beneficiary associated with this plan has an open private key pair. The private key utilized for signing is named as 'signature key' while the Public key is named as 'verification key.' Signer gives data to the hash function and creates a hash value and this hash value id directed to the signature algorithm. Hash value as well as signer's private key called signature key are then given to the signature algorithm which delivers the digital signature of the given hash [6].

At the receiver end, the verifier at that point provides a digital signature and the verification key is also known as the public key into the verification algorithm, and output is acquired. Verifier additionally runs a similar hash work on got data to produce a hash value [7]. At that point, this hash value and output from the verifier algorithm are looked at and compared. Furthermore, in light of the compared result, the legitimacy of the digital signature is checked.

Since Digital Signature is made by the private key of the sender and nobody else can have this key. Along these lines, the sender can't repudiate marking the data in the future. It ought to be seen that the hash of the data is utilized for signing the as opposed to signing the data itself. Utilizing hashes improves the productivity of the scheme.

6.4 IEEE 1609.2 Cryptography Mechanism

The IEEE 1609.2 [8] standard defines the use of the encryption techniques described below for the security of messages to be communicated.

For the sake of convenience in the following description, when communicating between two vehicles or a vehicle and an infrastructure, it is assumed that the entity transmitting the message is W_A and the entity receiving the message is W_B. We have use W_A and W_B as two communication nodes in description for IEEE 1609.2 cryptography mechanism at below.

6.4.1 Digital Signature

When W_A sends an authenticated message to W_B, W_A sends a message attaching a digital signature, which is a cryptographic checksum generated using its own private key, W_B can confirm that the message is sent from W_A by verifying the received message and the digital signature as the public key of W_A.

Such a digital signature is a technique particularly effective for security in communication with an entity that does not have a previous connection instance, such as broadcasting in a dynamically changing environment.

The 1609.2 standard backings the Elliptic Curve Digital Signature Algorithm (ECDSA) characterized as a computerized signature technology and alternatively remembers extra data for the digital signature.

6.4.2 Asymmetric Encryption Algorithm

The asymmetric encryption algorithm permits clients who don't share the secret key to impart safely. In this technique, a couple of a private key and a public key are utilized. A public key is revealed to a user taking an interest in the communication; however, a relating private key can be known uniquely to the owner.

When W_A encrypts a Message to be sent to W_B using the public key of W_B and then transmits it, W_B decrypts it using its corresponding private key. Generally, asymmetric cryptosystems have a disadvantage in that computation is more complicated than symmetric cryptosystems, and the method is used as a symmetric cryptosystem algorithm to increase efficiency. That is, W_A encodes a message utilizing a haphazardly produced secret key, encodes the secret key with the public key of W_B, and transmits the encoded message along with the message. In the asymmetric encryption procedure, just the short secret key is encoded, Symmetric encryption technology encrypts the entire message, which can increase efficiency.

6.4.3 Key Generation and Key Validity

ECDSA and ECIES, the private key and public key utilized are created by the definition in FIPS 196 − 3 * 9 as a key pair, and the utilization of a top-notch irregular number generator is requested.

The validity of the key pair is determined by IEEE Std. It is evaluated according to the criteria defined in 1363 − 2000 * 10.

6.4.4 Symmetric Encryption Algorithm

W_A may encrypt and transmit a message using a secret key, and W_B may decrypt the received encrypted message using the same cryptographic key. It also uses cryptographic keys to generate cryptographic checksum or message integrity check (MIC) to ensure authenticity and integrity.

The 1609.2 [8] standard uses AES-CCM (the propelled encryption standard-counter with cipher block chaining instrument) as the Symmetric algorithm, and AES-CCM is characterized in detail in NIST SP 800 − 38 C * 9 [9].

6.4.5 Implicit Certificate

Verifiable Certificate is a variation of a public-key certificate, which is a data structure containing the computerized signature esteem created by the certificate authority's private key for cryptographically and safely interfacing the identity data of the authentication holder and the public key.

Utilizing a digital certificate is known as the most ideal approach to build up identity in network information interchanges, and the authentication gives connectivity between the individual data and the public key. The key pair is utilized for the 7 digital signature required for key trade and approval of exchanges to set up secure communications. Along these lines, digital certificates are fundamental components in public key infrastructure (PKI).

Since the conventional public-key certificate contains a duplicate of the public key and the digital signature of the affirmation authority, the public key in the confirmation of the digital signature is utilized as in the individual recognized in the certificate knows the related private key and is the main party that realizes the private key expressly affirmed. It is along these lines called an explicit certificate. On account of a verifiable certificate, the public key is remade from the certificate, and the main party that realizes the related private key is called implicit regarding the individual recognized in the certificate.

In the case of an explicit certificate, the size of the certificate can be quite large, the infrastructure for the used key protection, the memory for storing and manipulating certificates, and the significant investment in securing the bandwidth required for repeatedly sending certificates to multiple objects need. On account of an implicit certificate, the encryption part of the certificate is extensively littler than the proven exact certificate. limited certificates are exceptionally helpful in situations where there are restricted assets where a great deal of memory or bandwidth isn't accessible. IEEE Std.

In Europe, the European Telecommunications Standards Institute (ETSI) has characterized the security prerequisites for the between vehicle communication dangers utilized in ITS through specialized report TR102893 [10] as follows:

- *Integrity [In]*

Objective		Functional security requirements
ID	Text	
In1	Data in ITS-S needs to be secured from unwanted or illegal change	Only allows authorized applications to modify or delete ITS-S security parameter and LDM information
		Only the above and authorized users can modify or remove service profile information
In2	Information transferred or received by a registered ITS user to be protected during transmission	An ITS-S shall implement one or more methods to enable it, if requested by an ITS user
In3	Management Information inside an ITS-S to be secured	In1 satisfies requirement
In4	Management Information transferred or received by ITS-S to be protected during transmission	Ln2 satisfies requirement

- *Availability [Av]*

Objective		Functional Security Requirements
ID	Descriptions	
Av1	Any malevolent action in the ITS-S ought not to confine access to and the activity of ITS services by approved clients	must have ability to detect attack patterns of denial of service

- *Confidentiality [Co]*

Objectives		Functional security requirements
ID	Description	
Co1	Parties without authorization (non-users) must not have information revealed to them	Assigning tags of restriction on information
		Restricted information to be encrypted
		Authentication from recipient
		Authentication needed from sender for recipient
Co2	Information in ITS-S to be kept away from someone without access	Only allow ITS application to access information
		Only authorized users to have access
Co3	Data on identity and administration capacities of an ITS client not to be uncovered to any outsider	Same as Co2
Co4	The executives Information moved to be shielded from individuals without approval	Only users can get management information
		Only get information from legitimate source
Co5	The management Information in ITS-S to be shielded from individuals without approval	entry restricted to only some users with authorization
		Provide a way for user to get access
Co6	By tracking communication between users, it should be impossible to locate them	User identity to be kept secret and not be added to location data in unlimited multicast address
		The above may be permissible for unicast or limited multicast address
		While transmitting should protect information
Co7	By tracking communication between users, it should be impossible to calculate the route taken	Shall have option to use multiple identifiers
		When used there will be no link between the identifiers

- *Accountability [Ac]*

Objective		Functional security requirements
ID	Descriptions	
Ac1	Changes in security application and parameters should be auditable	All changes and request for them in security shall be recorded

- *Authentication [Au]*

Objective		Functional security requirements
ID	Text	
Au1	Should not be possible for unauthorized users to alias themselves as legitimate users of ITS- S	Only authorized ITS-S get access to services
		Can validate identity of vehicle in the emergency
		In the emergency vehicle it should be possible to get temporary access to services
		ITS-S should be able to identify itself to emergency vehicle
		ITS-S will only be allowed to send massages if authorized in the situation
		Unauthorized messages to be ignored
Au2	Should not be workable for an ITS-S to get and process the management and configuration data from an unapproved client	In1 and Au2 satisfy requirement
Au3	Restricted ITS services like emergency warnings to be available only to authorize ITS users	Only currently authorized users will be allowed to transmit messages
		Authorizations will be either time limited or can be explicitly removed

6.5 Future of Automotive Cryptography

Automotive cybersecurity is a very important issue as this can be linked with the safety of the vehicles on the road. Lightweight cryptographic primitives could be the potential answer to ensure automotive cybersecurity. Customary cryptographic calculations, for example, public key foundation, elliptic curve cryptography, HASH functions, and symmetric key cryptography may not be applied legitimately in vehicular systems because of their high portability and dynamic system topology. Currently National Institute of Standards and Technology (NIST) has published a call for algorithms to be considered for lightweight cryptographic project. These lightweight cryptographic cipher proposals will go through a thorough three round evaluation phases over the next few years. These evaluation phases will consider both public analyses of the algorithms and the analyses provided by NIST internal committee.

6.6 Conclusion

In this article, we can get to know about the cryptography mechanism that can be used in automotive cyber security such as symmetric, asymmetric cryptography, we have get the overview of stream cipher, encryption and decryption of asymmetric key encryption an also digital signature. National Institute of Standards and Technology (NIST) is working on what's called post-quantum cryptography: public key cryptography that's resistant to quantum computers. So, we need ways of updating cryptography in vehicles such that when post-quantum options become available, they can be implemented in all vehicles rather than just new ones. The open acknowledgment for new innovation in vehicular systems must be guaranteed by advancing the security and protection of users.

Acknowledgements This research was funded by Woosong University Academic Research in 2021.

References

1. A.K. Jadoon, L. Wang, T. Li, M.A. Zia, *Lightweight Cryptographic Techniques for Automotive Cybersecurity*, vol. 2018. https://doi.org/10.1155/2018/1640167
2. M. Singh, Secure ID-based routing data communication in IoT. EAI Endorsed Trans Internet Things 18(6), 153566. ISSN 2424-1399. https://doi.org/10.4108/eai.15-1-2018.153566
3. B. Naik, D. Singh, A.B. Samaddar, H.-J. Lee, Security attacks on information-centric networking for healthcare system, in *19th International Conference on Advanced Communication Technology (ICACT)*, South Korea, Feb. 19–22, 2017, pp. 436–441
4. E. Schoch, F. Kargl, On the efficiency of secure beaconing in VANETs, in *3rd ACM Conference on Wireless Network Security (WiSec 2010), Proceedings*, March 2010
5. M.I. Salam, K.K.-H. Wong, H. Bartlett, L. Simpson, E. Dawson, J. Pieprzyk, Finding state collisions in the authenticated encryption stream cipher ACORN, in *Proceedings of the Australasian Computer Science Week Multiconference (ACSW'16)*. Association for Computing Machinery, New York, NY, USA, Article 36, pp. 1–10. https://doi.org/10.1145/2843043.2843353
6. M. Singh, I. Singh, IEEE E-Learning, securing intelligent transportation systems (2020). https://ieeexplore.ieee.org/courses/details/EDP587
7. M. Singh, Requirement engineering for intelligent vehicles at safety perspective. EAI Endorsed Transactions on Smart Cities, 2(6), (2017)
8. IEEE standard for wireless access in vehicular environments–security services for applications and management messages, in *IEEE Std 1609.2-2016 (Revision of IEEE Std 1609.2-2013)*, pp. 1–240, 1 March (2016). https://doi.org/10.1109/IEEESTD.2016.7426684
9. M. Dworkin, Recommendation for block cipher modes of operation: The CCM mode for authentication and confidentiality, National Institute of Standards and Technology (NIST), Technology Administration U.S. Department of Commerce, NIST Special Publication 800-38C-[Updated 2007]. https://nvlpubs.nist.gov/nistpubs/Legacy/SP/nistspecialpublication800-38c.pdf
10. ETSI TR 102 893 - V1.2.1 - Intelligent Transport Systems (ITS); Security; Threat, Vulnerability and Risk Analysis (TVRA) (2017). https://www.etsi.org/deliver/etsi_tr/102800_102899/102893/01.02.01_60/tr_102893v010201p.pdf

Chapter 7
Cybersecurity in Vehicular Communication

Madhusudan Singh

Abstract This chapter has presented the vehicle communication feature, for example, vehicle to vehicle (V2V) communication, vehicle to cloud, vehicle to system design. It presents the vehicular communication protocol explicitly committed short-range and remote access vehicular conditions and their cybersecurity challenges. Here some of the international standard cybersecurity standards for vehicular communication, for example, IEEE 1609, European Telecommunications Standards Institute (ETSI) security prerequisites within the vehicle to infrastructure environment have been discussed.

Keywords Intelligent vehicle · Autonomous vehicles · Information security

7.1 Overview

Vehicular Ad Hoc Network (VANET) are an expansion of the thoughts created in Mobile Ad Hoc Networks (MANETS), to the universe of vehicles. It is a thought that rotates around building a fluid network among close-by vehicles to permit the stream of significant data [1]. This system can be seen as a system taking into account communication between the following nodes.

- **Vehicle to Roadside Unit (V2RSU)**: Roadside units, are fixed units introduced on the street side in places like electric towers to communicate with different units in the network. It is likewise called vehicle to infrastructure (V2I) communication.
- **Vehicle to Cloud (V2C)**: Vehicle will send access to get the service information from the cloud servers.

M. Singh (✉)
School of Technology Studies, Endicott College of International Studies, Woosong University, Daejeon, Republic of Korea
e-mail: msingh@wsu.ac.kr

© The Author(s), under exclusive license to Springer Nature Singapore Pte Ltd. 2021   101
M. Singh, *Information Security of Intelligent Vehicles Communication*,
Studies in Computational Intelligence 978,
https://doi.org/10.1007/978-981-16-2217-5_7

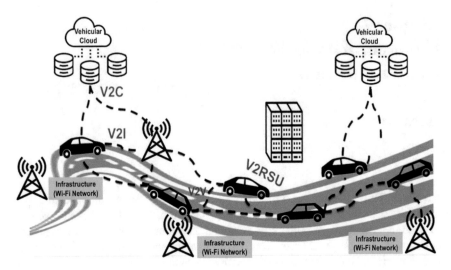

Fig. 7.1 Vehicular communication infrastructure

- **Vehicle to Vehicle (V2V)**: In-vehicle to vehicle (V2V) communication, vehicles share the information between, share traffic information, service information inside vehicles.

In Fig. 7.1 has presents, the vehicles wirelessly send information about themselves, the road, weather conditions, obstacles, accidents, and traffic information in general to other vehicles, the wi-fi network provided by the infrastructure and all this data is stored and shared in the vehicular cloud. This constant transportation network can empower applications that give security, versatility, and natural advantages to the transportation system. Vehicle-to-vehicle is represented as V2V, vehicle-to-infrastructure is V2I and vehicle to the side of the road unit is V2RSU), and vehicle to cloud is represented to as V2C [2].

To allow such a fluid network with continuously changing connections some sort of enhanced Wi-Fi technology is required. DSRCs or Dedicated Short Range Communications are the answer to this. DSRC are made to work in such fluid environments making them perfect for our need for quick time critical responses [3].

Communication techniques such as media access, data dissemination and routing etc. are used to provide steady communication. Figure 7.2 has represents the functionality of automotive communication techniques.

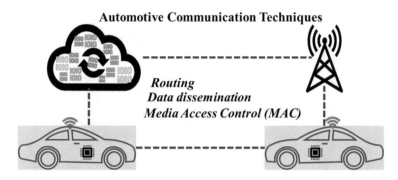

Fig. 7.2 Automotive communication techniques

They give the following functionality:

- *Media Access Control (MAC)*→ MAC support fast linking between units enhancing the reliability of the data being communicated over the short time constrained system.
- *Data dissemination*→ Data dissemination must be able to work in different types of network densities.
- *Routing*→ The routing in our system needs to be invariant to frequent topological changes.

The rest of the chapter is organized as follows. Section 7.2 describes the VANET characteristics with their challenges. Section 7.3 introduces Vehicular Communication protocols. Section 7.4 discusses the security process services in VANET communication. Section 7.5 conclude an overview of communication channels in vehicular environment. And future research directions.

7.2 VANET Characteristics

The following are some of the characteristic of VANETs which make it different for regular MANETs but also pose some unique challenges that needs to be addressed in the VANET communication [4].

Characteristic	Characteristic explanation	Challenge posed
High Mobility	Dynamic environment High relative speed	Frequent locality changes
Predictable and restricted mobility patterns:	Node changes decided by restricted set of rules Predictable	Increasing load on the channel
Rapid topology change:	High speed frequent changes in topology of the network	Connection irregularity

(continued)

(continued)

Characteristic	Characteristic explanation	Challenge posed
No power constraints	Vehicles are equipped with infinite power for all tasks	–
Localization	Use GPS for location with up to 1—5 m of accuracy	–
Abundant network nodes	Can have lots of nodes due to large number of vehicular traffic	Hidden terminals may cause packet losses
Hard delay constraints	Messages must be timely delivered as they have a validity time limit	Untimely irrelevant error messages

7.3 Vehicular Communication Protocols

It is conceivable to utilize any wireless networking system developed for IAV's. The most famous ones are Short Range Radio Technologies like Wi-Fi, Zigbee. Cell Technologies can likewise be utilized for VANETs like LTE. More current innovations like Visible Light Communication (VLC) are additionally expected to assume a significant job in the improvement of VANETs.

7.3.1 Dedicated Short Range Communication Technology

Dedicated Short-Range Communications (DSRC) [5] is an as of late created technology that furnishes short to medium range communication with stability and unwavering quality. In light of these characteristics, it is one of the insides focuses on exploration and applications in intelligent transportation systems with the need the security and portability the technology offers in a dynamic network environment. It might likewise be placed being used in applications other than the security ones in the event that quality as far as the nature of service is required. We can see an overview of dedicated short range communication in the Fig. 7.3.

- **Dedicated Short-Range Communication (DSRC) Process**
  Let us take a look in Fig. 7.4 at a model on how Dedicated Short-Range Communication (DSRC) works. The two vehicles in red have been engaged with a minor crash and are still. The green vehicle drawing nearer from the opposite side gets this information in a second and transfers it to a vehicle pushing toward it in the crash that appeared in dark. The same information may be received by the black vehicle from the two red vehicles but now the upcoming vehicle is aware of the fact that it is unable to enter the opposite lane and overtake the obstacle and that

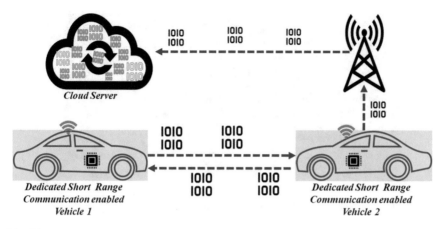

Fig. 7.3 An overview of dedicated short-range communication

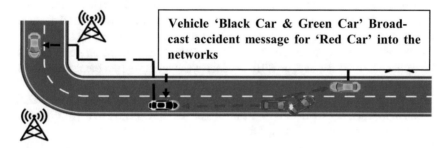

Fig. 7.4 Example of DSRC communication

it must come to a complete stop. All this information is conveyed to the blue vehicle following that has no optical view of the situation but nevertheless has been informed completely wirelessly.

- **Dedicated Short Range Communication Technical Characteristics**

  Dedicated Short Range Communication or DSRC is a communication convention or we can say protocol, which causes vehicles to communicate with other vehicles or systems with cellular networks. It is profoundly secure, rapid data transmission among vehicles, and encompassing infrastructure. The DSRC is a standard created by IEEE and known as 802.11p. The 802.11 standard utilizes carrier sense multiple access with crash avoidance whereby equipment tunes in to a channel for different clients (counting non 802.11 clients) before transmitting every packet. The 802.11p is a revision to the 802.11 standards which included wireless access in vehicular environment (WAVE). DSRC works in a 5.9 GHz band with low latency; the DSRC band assigned by the federal communication commission (FCC) is 5.850 MHz to 5.925 MHz.

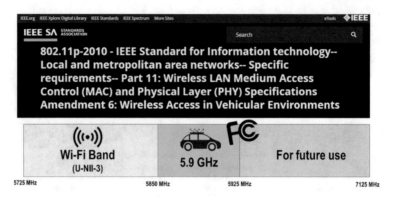

Fig. 7.5 Dedicated short range communication characteristics

The FCC has designated a range of 75 MHz in the 5.9 GHz band for such transportation security applications. The 5.9 here is part into seven 10 MHz channels. One channel is held for warning and alert messages and is known as the control channel. The others six are meant for all other required services while some on channels on the edges may be left idle for future use. These are therefore called service channels. These channels may also support different data exchange rate as per the requirements. We can find details in Fig. 7.5.

We can discover more details in the seven 10 MHz channels inside the 5.9 GHz band. This frequency works up to 1 km at 200 km/h. Channel 178 is utilized as control channel and channels 174, 176, 180, and 182 are utilized as service channels (SCH). Channel 172 and 184 are unused channels. It is essential to take note of that frequencies change in the United States, Japan, and Europe. In the United States, just the 5.9 GHz band is being considered for V2V safety communication of data interchange. This is the first Federal Communications Commission (FCC) range distribution that was chosen in 1999. The spectrum is distributed in the range 5.9 GHz band is 175 MHz. This takes into account a brief timeframe reaction that is under 50 m for each second. This is significant on the grounds that it identifies with response time as represents in Fig. 7.6.

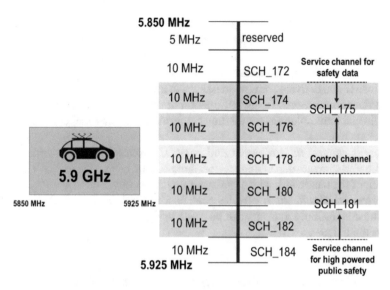

Fig. 7.6 Dedicated short range communication channel assignment

7.3.2 VANET Applications Enabled by DSRC

Due to all the aforementioned reasons and features DSRC technology is used in the below depict Fig. 7.7 and following type of applications:

Fig. 7.7 VANET applications enabled by DSRC

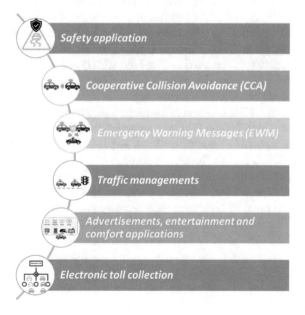

- **Safety application**: Vehicles delivers safety data in timely manner.
- **Cooperative Collision Avoidance (CCA)**: Sets trigger off when there is a possible collision between previously mentioned nodes, vehicles in our case.
- **Emergency Warning Messages (EWM)**: In case of any emergency event like an accident area, delivers warning messages to avoid danger.
- **Traffic managements**: Using information of position of vehicles/nodes manages traffic and finds solutions for diluting congestion, reducing fuel consumption and other such needs
- **Advertisements, entertainment and comfort applications**: To provide comfort and entertainment depending on the demands of users. These don't consume much of the bandwidth with security being the priority.
- **Electronic toll collection**: Vehicles don't need to stop to make payments and cause traffic congestion. Payments can be made using this application directly as cars pass the toll areas.

7.3.3 Wireless Access Vehicular Environment (WAVE) Communication Overview

As of late Intelligent Transportation Systems (ITS) have been broadly utilized for the productive activity of traffic and giving security and accommodation to vehicles users. WAVE is a technology standard established by the IEEE. It is a vehicle networking technology of V2V/V2I for public safety and ITS service. It is standardized in USA and Europe.

WAVE is a short-range wireless communication technology that delivers highway mobility and traffic conditions to a vehicle in a short time communication network. The IEEE 802.11a/g wireless LAN technology has been improved to meet the vehicle condition. Gives message exchange service through short-range wireless communication between a Road-Side Unit (RSU) and an On-Board Unit (OBU) or between vehicle-mounted terminals [6]. OFDM modem, between vehicle communication MAC and directing technology.

WAVE is standardized through a progression of IEEE 1609 standards documents, of which 1609.0 [7] is the architecture and service for multi-direct WAVE gadgets in a portable vehicle condition, and 1609.1 [7]. The asset director overseeing assets, 1609.2, depicts the administrations of uses and Management messages, 1609.3 [7] portrays networking administrations, 1609.4 multi-channel tasks. The guidelines for PHY and MAC layer are portrayed in IEEE802.11p. The structure of the WAVE layer and its relation to each standard are shown in Fig. 7.8.

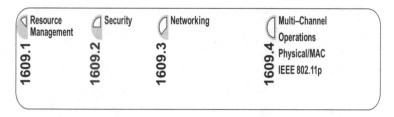

Fig. 7.8   IEEE 16.09 Wireless Access Vehicular Environment (WAVE)

7.4   Cybersecurity in WAVE Communication

WAVE communication technology is a safety service that earnestly transmits the risky data of the front streets and vehicles through between vehicle communication (V2V) and vehicle and infrastructure communication (V2I) in the driving condition of the vehicle to forestall resulting impact mishaps, (ITS), and the data transmitted through the correspondence is firmly identified with the security of the vehicle tenant and incorporates the contents of the individual privacy, so unapproved devices or individuals can get to the data or changing security threats as presents in Fig. 7.9.

Since WAVE communication is used for communication for ITS, safety related application is very sensitive to time, so it is necessary to keep the processing time and bandwidth overhead to a minimum, and since the object of communication may correspond to all vehicles on the road, the mechanism used for message authenticate needs to have a flexible and scalable structure. And the security mechanism not only protects the message used in communication from attack such as eavesdropping, information spoofing, alteration, replay, etc., Personal information must be protected to prevent exposure to unauthorized parties.

7.5   WAVE Communication Security Service Standards (IEEE 1609.2)

A lot of Standardization conventions of VANETs are occurring in the nations like U.S., Europe, and Japan, undifferentiated from their predominance in the automotive industry wildly in medium intelligent vehicles.

The IEEE 1609 WAVE (Wireless Access in Vehicular Environments) protocol runs on IEEE 802.11p WLAN, which works on 7 channel spectrums in a 5.9 GHz frequency bands shown in Fig. 7.10. The WAVE protocol stack is intended to outfit various channels in function and to likewise give security. There is a specialized board of trustees (technical committee) named, Vehicular Networks and Telematics Applications (VNTA) under the IEEE communication society. This advisory group advances specialized exercises in the field of Vehicular Networks.

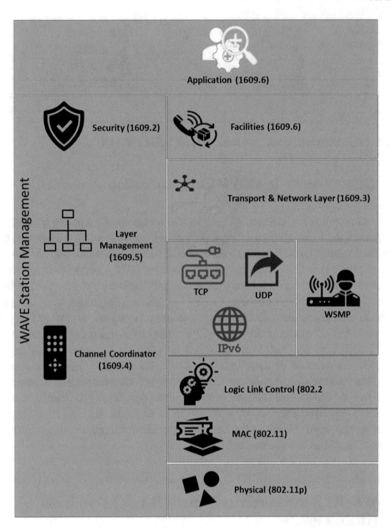

Fig. 7.9. Cybersecurity in WAVE communication

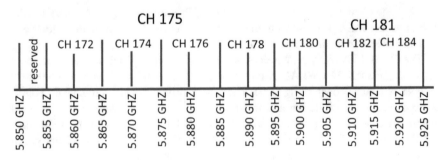

Fig. 7.10 WAVE channel allocation

In Europe, The European Telecommunications Standards Institute (ETSI) Intelligent Transportation System (ITS) Group 5 (G5) works on similar radio technology, having 5 stations in the 5.9 GHz frequency band. This protocol stack is an unpredictable hierarchy of such a large number of protocols joining to give a scope of basic services.

In Japan, ARIB STD-T109 also works on a similar technology, but it operates on a single frequency in the 700 MHz band. This protocol stack provides Time Division Multiple Access (TDMA), to divide the use between V2V communication and RSU services.

The IEEE 1609.2 [7] standard characterizes the security administration gave by the upper layer of the MAC for applications performed on the WAVE network stack and stack top. Security administrations comprise of Security Processing Services and Security Management Services as shown in Fig. 7.11 is a WAVE protocol stack indicating details of these security administrations? The security Processing Services gives a procedure to tie down communication to ensure vehicle information and WAVE Service Advertisements (WSAs). Security Management Services comprises of a Certificate Management Service and a Service Security Management Service. The Certificate Management Service is given by the Certificate Management Entity (CME) and gives data to the executive's administrations identified with the legitimacy of all things considered. The Service Security Management Service gives the private key and authentication-related data the executive's administration gave by PSSME

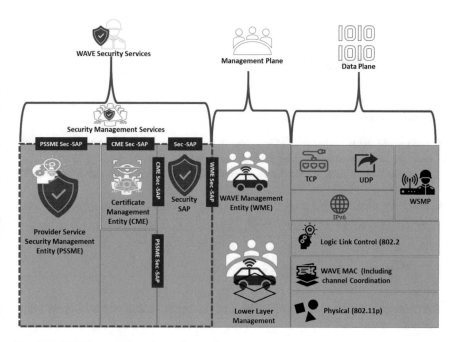

Fig. 7.11 WAVE protocol stack security services

Fig. 7.12 WAVE protection service

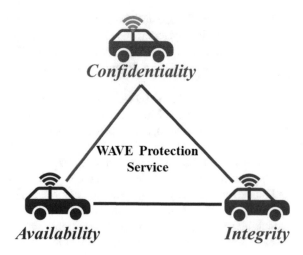

(Provider Service Security Management Entity) and used to transmit secure WSAs. In Fig. 7.11 has indicated the protocol stack of WAVE with security services.

WAVE communication Encryption method is utilized to ensure the important security necessities, for example, privacy, credibility, authenticity, and anonymity. In the IEEE1609.2 standard, encryption, digital signature (Digital Signature), and Hash Function are primarily utilized.

7.6 WAVE Protection Service

The IEEE 1609.2 standard defines the multiple functions to be provided through Security Services as represents in Fig. 7.12.

- **Confidentiality** Generated encrypted data is provided by security processing services and ensures the deciphering of encrypted data.
- **Authentication and Authorization** Generated signed data provided by security processing services and Verified signed data are guaranteed.
- **Integrity** Generated signed data provided by security processing services and Verified signed data are guaranteed.

7.7 Security Processing Services

The security processing services defined in the IEEE 1609.2 standard provide the following functions.

Fig. 7.13 Secure data
exchange

Secure Data Exchange

Fig. 7.14 Security
processing for security
management

Security processing for security management

7.7.1 Secure Data Exchange

Transforming unsecured Protocol Data Units (PDUs) into secured PDUs and transmitting them, receiving secured PDUs and converting them into unsecured PDUs, is a process of signing and encrypting PDUs before transmission (Encryption) and supports PDU decryption and verification at the time of reception. Signed WSAs It handles the Generate Secure WSA * 2 (Signed WSA) and the Signed WSA (Signed WSA) before transmission, signing the WSA before transmission, and the WSA when receiving it. Produced by the WME * 3 of the transmitting designed device and utilized by the WME of the getting device. The secure data exchange has represents in Fig. 7.13.

7.7.2 Security Processing for Security Management

To account the access of crypto materials, for example, private keys, public keys, and certificates declared in the security process; Generate a certificate request and procedure the reaction, approve the certificate revocation list (CRL). Figure 7.14 has represents an example of security processing for security management.

7.7.3 Private Key Associated Certificates

In the 1609.2 standards, the private key and related public key and certificate are put away inside security handling administrations and referenced utilizing Cryptomaterial Handle (CMH) *2. This system guarantees that the private key and public key or private key and certificate referenced by the CMH consistently structure a substantial pair: Related Services; Generate signed information; Verify signed information; Generate signed information; Decrypt encoded information; Generate signed WSA; Verify signed WSA on reception; Generate authentication demand; Verify response to certificate demand; Verify certificate repudiation list as shown Fig. 7.15 has presents.

- **Certificate Management Service**
  Certificate Management services manage the following information through the Certificate Management Entity (CME) so that security processing services can determine the authenticity of certificates and data received.
  Revocation and other status information for certificates whose corresponding private key is stored by security processing services. Revocation and information about certificates whose corresponding private key is not stored by security processing services. (Peer WAVE Security Service and Certificate Authority) as shown in Fig. 7.16.

Fig. 7.15 Private Key associated certificates

Private Key Associated Certificates

Fig. 7.16 Certificate management service

Certificate Management Service

- **Trust Anchor Used Certificate**

  The CME also manages information that can determine whether a received security PDU is a duplicate of a previously received PDU. As appeared in Figure, CME offers two Service Access Points (SAPs) for different entities to acquire and refresh security management data, one for CME-Sec-SAP and CME-SAP for use in different entities. Certificate management information is generated by the Certificate Authority (CA) * 2 and can be obtained in several different ways by an entity communicating with the WAVE security service. Figure 7.17 has presents the certificate-based trust anchor.

- **Service Security Management Services**

  Service Security Management Services are given through the Provider Service Security Management Entity (PSSME), which empowers security management data including Crypto material to be shared among WME and upper layer elements as shown in Fig. 7.18.

The requesting procedure is specified in IEEE Standard 1609.3. The detailed functions provided by PSSME are as follows.

Fig. 7.17 Trust anchor used certificate

Trust Anchor Used Certificate

Fig. 7.18 Service security management services

Service Security Management Services

- Assignment of Local Service Index for Security (LSI-S) allocation for security processing services to distinguish upper layer entities.
- Register the use of secure provider services so that processes outside the PSSME can apply for a WSA signing certificate.
- Communicate certificates and private keys with security processing services so that security processing services can sign WSA on behalf of WME.

For each known WSA signing certificate, PSSME allows rapid acceptance of repeated WSA by storing the most recently approved WSA signed with the corresponding certificate.

7.8   Conclusion

This article has presented the cybersecurity method and mechanism for vehicular communication protocol. Here we have described and shown the detailed information about dedicated short-range communication, remote access vehicular condition protocol and Its security challenges. This presentation likewise examined the potential solutions of communication protocol security issues and furthermore present brief perspectives on standard security protocols, for example, IEEE 1602, and so on. At that point, the European Telecommunications Standards Institute (ETSI) gives the Security prerequisite details in Vehicular communication inside Intelligent Transportation System (ITS).

Acknowledgements This research was funded by Woosong University Academic Research in 2021.

References

1. S. Zeadally, J. Guerrero, J. Contreras, A tutorial survey on vehicle-to-vehicle communications. Telecommun. Syst. **73**, 469–489 (2020)
2. I.A. Abbasi, K.A. Shahid, A review of vehicle-to-vehicle communication protocols for VANETs in the urban environment. Future Internet **10**, 14 (2018)
3. Z. Junping, W. Fei-Yue, W. Kunfeng, L. Wei-Hua, X. Xin, C. Cheng, Data-driven intelligent transportation systems: survey. IEEE Trans. Intell. Transp. Syst. **12**(4), 1624–1639 (2011)
4. H. Hasrouny, A.E. Samhat, C. Bassil, A. Laouiti, VANet security challenges and solutions: a survey. Veh. Commun. **7**, 7–20 (2017)
5. https://standards.ieee.org/standard/1609_3-2016.html
6. https://wiki.campllc.org/display/SCP/Test+Vectors
7. https://portal.etsi.org/webapp/workprogram/Report_WorkItem.asp?WKI_ID=54222

Chapter 8
In-Vehicle Cyber Security

Madhusudan Singh ⓘ

Abstract Recently, the utilization of electronic control units (ECUs) from the well-being of vehicles to infotainment has been incredibly expanded. Thus, vehicles are progressively advancing as smart vehicles or associated vehicles in the basic mechanical system and self-driving vehicles are relied upon to develop sooner rather than later. This presentation will introduce the in-vehicle review and their security challenges and furthermore conceivable solutions.

Keywords V2V communication · Electronic control unit · In-vehicle cyber security · Information security

8.1 Introduction

The evolution of such a car increases the dependence on information sharing among in-vehicle ECUs, communication within the vehicle, and increases the connection with the outside, resulting in the opening of a new attack surface [1]. In order to defend against this, a strong security solution is required in some form.

C. Miller and C. Valasek recently identified 20 categories of vehicles launched in 2014 and 2015 and identified seven categories of Remote Attack Surface, recognizing the severity of vehicle hacking. The success of vehicle hacking depends on three main categories: Remote Attack Surface, Cyber physical Features, and In-Vehicle Network Architecture [2].

According to Intel, "Security of complex systems such as smart cars requires a collaborative, holistic approach with participation of the supply chain and a wide range of ecosystems, and effective security cannot be achieved by responding to the threats or attacks of individual components, unlike traditional computer systems, it makes it more difficult to protect vehicle systems because they can attack vehicles in both the cyber world and the physical world." From this point of view, security for

M. Singh (✉)
School of Technology Studies, Endicott College of International Studies, Woosong University, Daejeon, Republic of Korea
e-mail: msingh@wsu.ac.kr

© The Author(s), under exclusive license to Springer Nature Singapore Pte Ltd. 2021   117
M. Singh, *Information Security of Intelligent Vehicles Communication*,
Studies in Computational Intelligence 978,
https://doi.org/10.1007/978-981-16-2217-5_8

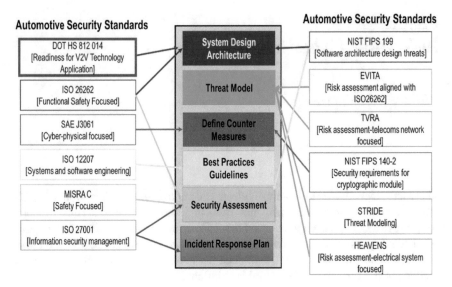

Fig. 8.1 Six Common cyber security practices for automotive technology

vehicle systems needs to refer to the Framework * 1 for the Cyber-Physical System defined by National Institute of Standards and Technology (NIST).

The rest of the chapter is organized as follows. Section 8.2 provides an overview of automotive cyber security practices; Sect. 8.3 introduces requirements of security in-vehicle. Section 8.4 in-vehicle network security. Section 8.5 summarize the in-vehicle security and future research directions.

8.2 Automotive Cyber Security Practices

We have presented in Fig. 8.1. Six common practices related to vehicular cyber security, these practices are defined and described in detail by specific automotive security standards. In Fig. 8.4 has indicated, which standards correlate to what practices. To understand better the existing practices, one must familiarize himself with the standards that guide them [3, 4].

8.3 In-Vehicle Security Requirements

In this part, we are going to discuss about the In-vehicle cyber security requirements. In-vehicle cyber security requirements are based on vehicles connectivity, access, Electronic Control Unit (ECU), Control Area Network (CAN) data platform and communication environment [5]. The security requirements are described in Fig. 8.2.

E-Safety Authentication and Integrity	Authenticity and Integrity of events which rely on critical information shall be assured based on some factors such as origination of event, subject matter and time. Any type of Falsification, manipulation or reduplication shall be notified.
ECU Installation Authentication and Integrity	Genuine ECU and authentic installation or any replacement shall be done in the vehicle. Newly Security algorithms and important information uploaded in the vehicle shall be protected from adversaries.
Privacy	The private data saved in the vehicle or messages send during communication of vehicles shall be adhered with privacy policy. For example Messages bearing a link shall be bounded by the vehicle applications
Confidentiality	The software, ECU's, newly enabled configurations or security certifications shall be kept confidential.
Access Control	The vehicle's data or resources and operations shall be only accessible by legitimate users and not by the adversaries.
Reliable ECU Platform	The running software's integrity and authentication shall be assured. A non-trusted configuration shall not run any modified platform.
Secure Data Storage	The data stored in the vehicle shall hold integrity, confidentiality, privacy and functionalities shall be only used by application's legitimate users ensuring access control.
Security functionality Intervention	The availability of the CPU's, Bus system, RAMs shall not oppose the impact of security service operations.
Secure Run Environment	Any harm to ECUs shall not lead to system wide attacks, mainly concerning the e-safety applications. The effects of the successful ECU attacks shall moderately impact the reliable portions of the platform.

Fig. 8.2 The security requirements for in-vehicles

8.4 Vehicle Security Requirements Projects

The security for the vehicle system includes Secure Processing for in-vehicle ECU, Secure Network for in-vehicle network, Secure Car Access and Secure Gateway and Secure Interface are required [6]. In Fig. 8.3 has shown the position of security requirements in a vehicle.

Various Vehicle Information Security projects such as Secure Hardware Encryption (SHE), Hardware Security Modules (HSM), E-safety Vehicle Intrusion protected Application (EVITA) are supported by European Transportation Association (ETA) and currently they are active on these projects. These projects focus on eliciting security requirements and security objectives for In-Vehicle Security. The elementary security objective is to meet the in-vehicle security specified goals and objectives

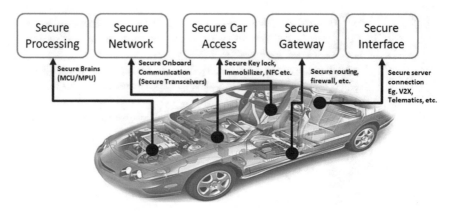

Fig. 8.3 In-vehicle security model

Media Oriented Systems Transport (MOST)
Ethernet AVB (Audio Video Bridging)
Ethernet TSN (Time – Sensitive Networking)

FlexRay
Brake by Wire System

Controller Area
Network (CAN)

Multifunction
Keyless System

ECU

Ethernet

Local Interconnect
Network (LIN)

Fig. 8.4 In-vehicle security

while sustaining the functional performance of vehicles development and vehicle's security services.

The key objective of the EVITA projects is to prevent unauthorized access inside the car from adversaries during vehicle communication in the network. It is primary concern to protect any illegitimate modification of the Vehicle application. During vehicle communication, vehicles must disclose the information of operation. In-vehicle security also comprises of protection of intellectual property rights of the vehicle parts, the suppliers, and the developers.

8.5 Isolation and Virtualization

Although various encryption and authentication technologies are used as a defense method for vehicle security, the most basic method is to prevent the internal system and the network from being exposed to the outside. This requires the utilization of a different network between vehicle ECUs that share data, for example, the utilization of a passage that isolates the vehicle's distributed internal system network all things considered. It is likewise important to virtualize the vehicle data so as to forestall direct access to the vehicle ECU for gaining vehicle data from the external network and to forestall change of the vehicle data.

Isolation is formed by grouping vehicle ECUs belonging to the same domain to form a network, connecting the domains to each other through a domain gateway, and connecting the whole network to an external network through a vehicle communication gateway (VCG) [7]. In this regard, it is considered necessary to study the contents necessary for grouping considering the security and function of the vehicle, not the function-oriented domain (chassis domain, body domain, engine domain, etc.). The requirements for security-related functions of the Domain Gateway and the VCG and the criteria for efficiently partitioning each other's roles should also be standardized. For the VCG, ISO TC204 defines the role of the ITS station gateway and the connection between the portable device (Nomadic Device) and ISO TC22 SC31, which deals with the data communication of the road vehicle, in order to establish the VCG standard for sensitive vehicle information access.

Virtualization prevents other devices other than the authorized vehicle ECU from directly accessing specific ECUs and accessing the vehicle information held by the ECUs, and requires a separate secured CPU [8], storage device, and management system for encapsulation Do. Therefore, it is necessary to define the requirements for the definition and management of the virtualized vehicle information architecture, the criteria for classifying the vehicle information to be virtualized, and the communication protocol for the vehicle ECU and the virtualized vehicle information management system need.

8.6 In-Vehicle Network Security

For the in-vehicle Network Security, the data and information are exchanged using the same communication bus. The communication bus utilizes CAN (Controller Area Network), FlexRay, LIN, etc. According to the recent research on Automotive Ethernet, CAN was first developed by BOSCH and later was launched as an international standard for using internal serial communication buses in ISO. CAN message is carried to all the other ECUs linked to the CAN bus in broadcast mode. The ECU is sent in the DATA frame of one message and is authenticated by another message and this authentication is not certified and not supported by an authenticated encryption protocol as there is no DATA field [9] as shown in Fig. 8.4.

8.6.1 Electronic Control Unit [ECU] Protection

Although there are many aspects of vehicle system security requirements, it is necessary to consider safety as a top priority. From this point of view, it is necessary to protect the ECUs inside the vehicle from cyberattacks so that the designed functions and performance can operate properly, we can see in Fig. 8.5.

There are standards for automobile security that incorporate cyber security concepts into existing development processes, standards for vehicle security architecture, and standards for HSM. However, most of them are definitions at higher level, so further research and development is required.

In the case of the current HSM, the built-in encryption and digital signature technologies are suitable modules for current ITS vehicle communication, but the scalability is very weak when the more secure encryption technology is developed in the future. Therefore, it is necessary to develop and standardize the HSM architecture that can be easily updated when security vulnerabilities are detected, considering scalability.

And as more smart cars get smarter, and more connected cars are deployed, cyberattacks on cars will be more exposed to attacks and more vulnerabilities will be discovered. If the existing automobile maintenance procedure is to be countered, it will have a serious effect on the safety of the vehicle. Therefore, it is necessary to develop a method of safely, easily and quickly updating the ECU or the security

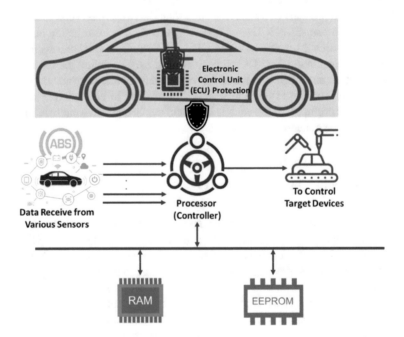

Fig. 8.5 Electronic control unit protection (ECU)

program of the vehicle. OTA (Over the Air) is considered as a solution to this problem. We think that it is necessary to design the ECU architecture that links this with the vehicle security, and to update the security code including the ECU algorithm as well as the ECU firmware.

8.6.2   Hardware Security Component

The hardware security component is fashioned to hand over optimal level of security, service, and performance. The security and performance requirements of signing message and V2X communication verification are satisfied by its highly effective asymmetric encryption engine. Hardware security component consists of Encryption BB (Building Block) where all cryptographic hardware functions are executed. It consists of an Asymmetric Cryptographic Engine (ECC-256-GF(p)), and an AES based hash function known as WHIRLPOOL. It also consists of AES-128 encryption/decryption engine, a pseudo random number, AES-PRNG, a 64 bit monotonically increasing counter and a logic BB that associates EVITA hardware with the ECU application core. An HSM internal CPU is also present to deal with all logics and non-time-critical cryptographic functionality. Further, the hardware security component consists of a 64 KB RAM, 10 KB ROM, a 32 KB non-volatile memory to store keys, security credentials, and counter values and a secure EVITA hardware interface called as HW-API which is used to access all the security functionalities for the software and application CPU.

8.7   Summary

Security in CAN BUS can be achieved by using several authentication protocols and authentication tags such as Message Authentication Codes (MAC). To improve CAN's limitation on the Automotive Vehicle, Automotive Ethernet is being analyzed as an option to the new Network Architecture and also proposed to develop it as a standard in the in ISO TC22 SC31.

Acknowledgements This research was funded by Woosong University Academic Research in 2021.

References

1. J.H. Kim, S.-H. Seo, N.-T. Hai, B.M. Cheon, Y.S. Lee, J.W. Jeon, Gateway framework for in-vehicle networks based on CAN, FlexRay, and Ethernet. IEEE Trans Vehic Technol 64(10) (2010). https://doi.org/10.1109/TVT.2014.2371470
2. D.A. Brown, G. Cooper, I. Gilvarry, A. Ranjan, A. Totourian, R. Venugopalan, D. Wheelerr, M. Zhao, *Automotive Security Best Practices*. White Paper: McAfee Intel Security, US (2015)
3. M. Cheah, S.A. Shaikh, J. Bryans, P. Wooderson, Building an automotive security assurance case using systematic security evaluations. Comput. Secur. **77**, 360–379 (2018). ISSN 0167-4048. https://doi.org/10.1016/j.cose.2018.04.008
4. V. Rajarajan, K. Nedungadi, B. Bhatia, C. Kiernan, A. Ganesan, J. Johnston, L. Gallagher, K. Hodge, T. Martino, C. Rohwer, A. Hayes, A.B. Hall, A. Zimnicks, M. MacMahon, *User Interface for Managing Multiple Network Resources*. US 7689921B2, US Patent (2000)
5. M.A. Rahman, Q. Duan, E. Al-Shaer, Energy efficient navigation management for hybrid electric vehicles on highways, in *Proceedings of the ACM/IEEE 4th International Conference on Cyber-Physical Systems (ICCPS'13)* (ACM, New York, USA, 2013), pp. 21–30. https://doi.org/10.1145/2502524.2502528
6. S. Burnett, S. Paine, *RSA Security's Official Guide to Cryptography* (RSA Press Book, McGraw-Hill Publication, 2001). https://doi.org/10.1036/0072192259. https://www.scribd.com/document/325732034/RSA-Security-Official-Guide-to-Cryptography
7. J. Joy, S. Samuel, V.S. Vinu, *White Paper: Gateway Architecture for Secured Connectivity and in Vehicle Communication* (Tata Elxsi, 2015)
8. C. Bernardini, M.R. Asghar, B. Crispo, Security and privacy in vehicular communications: challenges and opportunities. Vehic. Commun. **10**, 13–28 (2017). ISSN 2214-2096. https://doi.org/10.1016/j.vehcom.2017.10.002
9. A. Muneeswaram, Automotive diagnostics communication protocol analysis KWP2000, CAN and UDS. IOSR J. Electron. Commun. Eng. (IoSR-JECE) **10**(1), 20–31 (2015). e-ISSN: 2278-2834, p-ISSN: 2278-8735

Chapter 9
Vehicle to Vehicle Communication Protection

Madhusudan Singh ⓘ

Abstract In this article we have described basics of security, Security impact and provide and, overview on security areas and their vulnerable places such as Computer security, Network Security, and Cyber security. These security features must not be limited to protecting confidential information, but also needs to address safety critical systems such as brake, accelerator or steering etc. In this presentation, we have talk about the essentials of the vehicle-to-vehicle communication features and their difficulties and talk about the review on potential solutions. We have additionally referenced overall cyber security ventures summary.

Keywords V2V communication · Information security · Secure credentials management system · Conflict over spectrum

9.1 Introduction

V2V is an abbreviation for Vehicle to Vehicle, which is a automotive innovation that permits vehicles to communicate with one another. Major automakers are chipping away at this technology, this incorporates GM, BMW, Audi, Honda, and the Car-to-Car communication consortium. In 2006, General Motors had exhibited this system utilizing Cadillac. V2V is additionally in some cases named as VANETs (Vehicular Ad Hoc Networks). It is a variety of well-known portable system Mobile Ad-hoc NETworks (MANETs) [1], with vehicles going about as node. In 2001, it was referenced in distribution, that vehicles can shape ad hic networks, which would then be able to help in evading mishaps and crash.

There has been a lot of research and on-road tests going on here, VANETs are being applied in a variety of applications, e.g., security, route, and so forth. The US Department of Transportation proposed rules to make V2V communication capacities, mandatory for light-obligation vehicles [2].

M. Singh (✉)
School of Technology Studies, Endicott College of International Studies, Woosong University, Daejeon, Republic of Korea
e-mail: msingh@wsu.ac.kr

© The Author(s), under exclusive license to Springer Nature Singapore Pte Ltd. 2021   125
M. Singh, *Information Security of Intelligent Vehicles Communication*,
Studies in Computational Intelligence 978,
https://doi.org/10.1007/978-981-16-2217-5_9

In order to avoid any kind of mishap and for ease of driving, information about the driving conditions of the vehicle must be communicated with all the nearby vehicles. Thus, one logical way of information exchange is to multicast the information of the vehicles. In this manner, all of the vehicles which are operational near to the transmitting vehicle, will receive the multicast packet simultaneously. This type of communication is a type of point-to-multipoint communication.

The rest of the chapter is organized as follows. Section 9.2 provides an overview of v2v communication security design. Section 9.3 introduces the Intelligent vehicles communication elements. Section 9.4 discusses Vehicle 2 Vehicle Communication Utilizations (Application). Section 9.5 summarized available V2V Communication application and future research directions.

9.2   V2V Communication Security System Design (DOT HS 812 014)

The initial deployment model as well as the full deployment model of a Secure Credential Management System (SCMS). The SCMS focuses on communications and activities within the vehicular network. No human judgment is involved in creating, granting, or revoking digital certificates. The functions are performed automatically by processors in the various V2V components, including the On-Board Equipment (OBE) in the vehicle. The role of personnel within the SCMS is to manage the overall system; protect and maintain the computer hardware and facilities; update software and hardware; and address unanticipated issues. These SCMS operating functions fall into two categories: pseudonym functions and bootstrap functions. An SCMS has shown in Fig. 9.1.

- *Pseudonym Functions (PF)*: The vehicular network infrastructure provides short term digital certificates for on-board vehicles in security design. Model and that digital certificates used for authentication and validation for safety messages that broadcast between the vehicles during vehicle to vehicle (V2V) communication. That short-term digital signature only represents the credential of vehicle that involve in V2V communication without containing any private information of the vehicles.
- *Linkage Authority (LA)*: The linkage authority (LA) is mostly operated for Registration Authority which provide values for registration authority and pseudonym CA on their request. The linkage values are calculating a certificate ID and provide the connectivity of short-term certificates with specific device for ease of revocation in the event of misbehavior. In Secure Credential Management System, the LA has come in pairs, which refer to Linkage Authority 1 and Linkage Authority 2 as shown in Fig. 9.1.
- *Location Obscurer Proxy (LOP)*: The Location Obscurer Proxy (LOP) provide the unclear (Obscure) location to On-Board Equipment (OBE) because the OBE seeking to communicate with the SCMS functions. That means LOP provide the

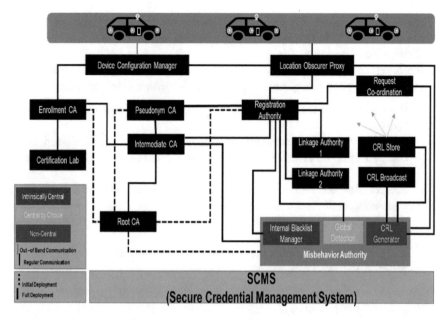

Fig. 9.1 An overview of the secure credential management system

proxy (hidden or unclear obscure) geographical location of particular vehicle. Whenever any vehicle communicates with OBE to SCMS, all communications must route via the LOP. During the full deployment, the LOP may shuffle misbehavior reports that are transmit by OBEs to the Misbehavior Authority (MA). The LOP not increase or reduce the security of networks, it's just increase the privacy of participants

- **Misbehavior Authority (MA):** The Misbehavior Authority (MA) is authorized to process misbehavior reports and generate with publish the certificate revocation list. It received the information from the pseudonym Certificate Authority, Registration Authority, and Linkage Authorities.to create entries to the Certificate of Revocation list through the CRL Generator. The MA behave like central function that eventually perform global misbehavior detection with the investigation of misbehavior level in the system. The MA is an internal SCMS functions which is responsible to investigate the malfunction, and message authenticity within the system.
- **Pseudonym Certificate Authority (PCA):** The pseudonym Certificate Authority (PCA) provide the short-term certificates to ensure trust in the system and that certificate validity was fixed at five minutes in earlier design but now it is depending on variable of length of time making them less predictable and harder track even though the certificate life in order of minutes. The PCA issued certificate used to authenticate messages from a device. Additionally, It has issued 170

pseudonym certificate for any kind of misbehavior where other authority collaborates with the MA, RA, and LA to find linkage values to add on the Certificate Revocation List.

- *Registration Authority*: Before the PCA, the Registration Authority (RA) expended the compulsory key. It performs with below mentioned 6 functions:
  - RA receives to issue the certificate requests from the On-Board Equipment's (OBEs)
  - RA request and receives the linkage values from the Linkage Authorities.
  - RA transmits the certificate request to the Pseudonym Certificate Authority (PCA)
  - RA shuffles the requests from multiple OBEs to protect the PCA from correlating certificate IDs with users.
  - RA acts as the final conduit to batching short term certificates for distribution to the OBE.
  - RA creates and maintains a blacklist of enrollment certificates so to reject certificate renewal requests from revoked OBEs.

- *Request Coordination (RC)*: The Request Coordination (RC) provide the protection of OBE from the multiple batches of certificates which received from different Request Authorities such as LA, PCA, MA, RA etc. This RC function make the coordination with the Registration Authorities (RA) to make sure that response of certificate request has completed within given time without any duplication. This function has important only if SCMS has more than one RA.
- *Root Certificate Authority (RCA)*: The Root Certificate Authority (RCA) is the centralized trusted authority and boss of all other Certificate Authorities (CA), It also known as "Center of Trust" of the system. The RCA provide the certificate to other CAs in a hierarchical form and authenticate within system. The other users and functions can trust, if they know, certificate of other functions has issued by RCA. The RCA produces a self-attested certificate to use the out of band communications. It builds the trust due to using common trust point between CAs and verified between ad-hoc networks. Due to it has catastrophic behavior of RCAs, its works separately and offline environment.
- *Secure Credential Management System (SCMS) Manager*: The Secure Credential Management System (SCMS) is creating milestone for the entire connected vehicle environment in respect of policies, and technical standards of the security issues. And solutions which consider by the automotive industry. The SCMS manager must ensures to the consistency and standardization of technical specifications, standard operating procedures, and other automotive industry wide practices such as auditing security implementation etc.

9.3 Conflict Over the Spectrum

The V2V technology has a genuine danger from satellite TV and other tech firms. They need to get a major piece of radio spectrum from the data transfer capacity which is as of now V2V's property. They need to utilize these frequencies to give high-speed internet service providers. V2V's portion of the radio spectrum was put aside by the administration of the USA in 1999, however, has not been utilized yet [3].

The National Transportation Safety Board is supporting the side of the automotive vehicle industry, while Federal Communications Commission (FCC) is in favor of tech organizations. These organizations are asserting that, V2V communication for self-governing vehicles is superfluous. The US automotive industry is prepared to share the spectrum if V2V service doesn't get intruded. Also, FCC is wanting to test a few sharing plans. There are still nations that do not have spectrum held for V2V communication, so vehicles should experience the major bad effects of different interchanges [4]. The spectra saved for V2V communication in certain nations are as per the following:

Locale	Spectrum
USA	5.855–5.905 GHz
Europe	5.855–5.925 GHz
Japan	5.770–5.850 GHz; 715–725 MHz
Australia	None

9.4 Benefits of Vehicle-To-Vehicle Communication

Vehicle-to-Vehicle communication are essentially a remote system, where various vehicle speaks with one another, containing data about what they are doing. This communication can move reports like area, speed, course of movement, loss of strength, and slowing down. This technology utilizes Dedicated Short-Range Communications (DSRC), a standard proposed by bodies like ISO and FCC. Some of the time, it is likewise named as Wi-Fi network, in light of the fact that one of the potential frequencies is 5.9 GHz, which is utilized for Wi-Fi services, at the same time, it's smarter to call it "Wi-Fi-like". It has a scope of 300 m or around 10 s at roadway speeds.

V2V is a mesh network, which implies that each node (vehicle, smart traffic signals, and so on) in the system can send, catch, and retransmit signals. Five to ten hops on the system can get the traffic conditions a mile ahead. This gives sufficient opportunity to a driver to take his choices.

Technology which enables communication between stationary devices (like traffic signals) and vehicles is termed as V2I, or Vehicle to Infrastructure. V2I is sometimes

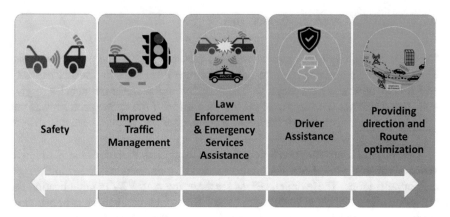

Fig. 9.2 Benefits of vehicle-to-vehicle communication

merged in V2V to avoid these many three letter words. Some automobile company use terms like Car-to-X instead of V2V, here X can be anything. Other terms like "Internet of Cars", "Connected Cars" and "Talking Cars" have also become very popular. And, out of all of them, V2V is the most used phrase.

In December 2016, the US Department of Transportation trained that, it is obligatory to have V2V communication technology in all the new vehicles by 2023. Also, after this affirmation, automakers are starting to incorporate V2V innovation in their vehicles. A case of this is 2017 Cadillac CTS, which has DSRC communication devices that can share data like GPS location, speed and direction routes [5]. Some of the benefits of using V2V communication technology in vehicles are listed below as sown in Fig. 9.2.

Each benefit has described at below section.

- **Safety**

  As indicated by an investigation, directed at Stanford, 93% of vehicle mishaps are caused because of human blunder. Along these lines, with self-driving vehicles on the streets, this blunder can be expelled by communicating with the environmental factors. Which thus will decrease street mishaps. As we can see at Fig. 9.3. Vehicles are communicating their speed, location and traffic information with each other's according to information they have manage their safe driving process such as accelerator, brake etc.

Fig. 9.3 Safety process
within V2V communication

- **Improved Traffic Management**
  V2V communication provides real-time information about the status of the traffic at various routes, So, with this information drivers can be guided to avoid heavy traffic routes as shown in Fig. 9.4.
- **Law Enforcement and Emergency Services Assistance**
  V2V communications can be quite useful for emergency services. If in case, any accident occurs or fire catches up, then other vehicles can be instructed to avoid that route and provide the passage for Ambulance and Fire services. Police can also use V2V to warn careless drivers. As we can see in Fig. 9.5, accident information has communicated by vehicles to vehicle (Emergency vehicle) and emergency service authorities act as soon as possible according to law enforcement, so traffic flow not interrupt for long time an it's manage as soon as possible.
- **Driver Assistance**
  V2V and in-vehicle communications can improve the driving experience in almost every aspect of driving. Notifying the driver of traffic conditions, finding spaces

Fig. 9.4 Improved traffic management

Fig. 9.5 Accident information received by emergency vehicle (police)

Fig. 9.6 Driver assistance

for parking, etc, are some of the ways in which driving assistance can be provided to the driver. Figure 9.6 has image has assist to driver according to sign board driver drive his vehicle carefully.

- **Providing direction and Route optimization**

  This vehicle communication can provide the direction of the routes to the drivers. Destination location on Map and ways for route optimization can be communicated through this.

  If there will be large number of vehicles having the V2V technology, then the information collected from one vehicle can be used to provide journey optimization tips to other vehicles. In other words, drivers would not have to rely only on GPS, which only provides route optimization on the basis of distance and speed but will be familiar to the condition of the routes of the present situation. This will result in an improved customer service to the drivers. When the best route gets congested, then the system can alert the driver to change the route [6].

  In the coming future, V2V communication will bring a lot of changes in the automobile industry, with improvements in safety as the main purpose behind its implementation. Apart from safety, there are tons of other reasons why, drivers will be benefited by V2V communication technology as shows in Fig. 9.7.

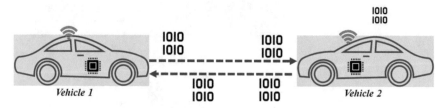

Fig. 9.7 Providing direction and route optimization based on communication way

9.5 Intelligent Vehicles Communication Element

The USDOT is right now investigating ways, to empower vehicles to speak with one another or with the infrastructure around them, with the assistance of wireless technology, in a joint effort with, some of the world's biggest automobile producers.

Dedicated Short Range Communication (DSRC) will be utilized for a security-related system of vehicles. DSRC is a lot of like WiFi. It is a secure, quick, and solid innovation. Applications other than security might be based on some unique wireless technology. Vehicles will have the option to speak with one another, and with the infrastructural devices to impart significant data to one another.

While sharing the data, the secrecy of the driver or vehicle is kept up, which forestalls vehicle tracking and furthermore meddling with the system. Figure 9.8 has presents an example of vehicle to vehicle (V2V) communications.

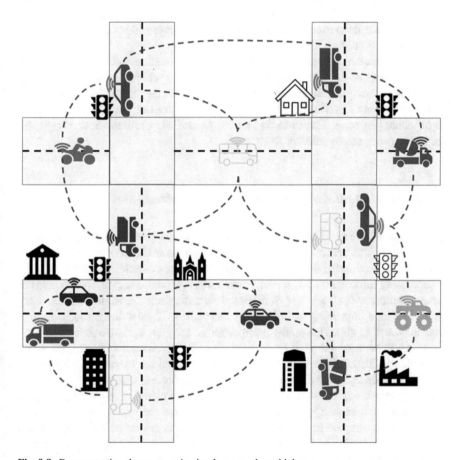

Fig. 9.8 Demonstrating the communication between the vehicles

Fig. 9.9 Intelligent vehicle
improve safety

9.5.1 Intelligent Vehicles Improve Safety

V2V safety applications will allow drivers to know about all the dangers around themselves, which they can't even see. V2I will enable drivers to know about the area, or the conditions, in which they are driving, for example, they will be alerted if they are entering school zone, or if the traffic light is about to change [7]. Crucial works have been done to ensure that, these driver warnings will not distract the drivers as shown in Fig. 9.9.

Communications between the vehicles and infrastructures will generate tons of data, which can be used by transportation managers to analyze them and come with ways to decrease traffic and make roads safer. The data can be manipulated to bring up the solutions of millions of our daily problems. For example, apps that can help us to locate bus, train, etc., or to find parking spaces. Overall, this technology will further improve the standard of living.

9.5.2 Intelligent Autonomous Vehicle Environment

The US government had contributed 28.2% of the total Greenhouse emissions in the year or 2018. The vehicle-to-vehicle communications make ease to drivers to take better decision while driving according to situation. For example, a particular vehicle broadcasting information about traffic scenario in particular region, so it is helps to other vehicles will take better route based on information, avoiding unnecessary stops, and also improving fuel efficiency their plans. It's also helps to getting real time information about the public transportation, travelers optimize their plans.

As we know now a days, a lot of research going on autonomous vehicles which are enabled with automated eco-driving. This automated eco-driving techniques will help us to reduce the environment degradation. It has set practices to reduce the fuel consumption in the vehicle with changing any mechanics of the vehicle. This feature can be achieved by optimization of speed and the engine power in use to maintain adequate power output and also based on acceleration optimization and deceleration behavior of the drivers. The autonomous eco driving. It can observe that human drivers' actions such as wrong acceleration and deceleration techniques, either accelerate too fast or they pushed brake too forcefully. This type actions make

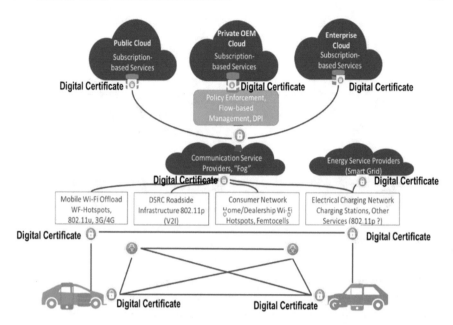

Fig. 9.10 Intelligent autonomous vehicle environment

wastage of energy and fuel consumption. If the vehicles will employ automated eco-driving technique, then it is assumption that 10–25% energy consumption of vehicles will be reduced on the road. Figure 9.10 has represents an example of Intelligent autonomous vehicle environment.

9.6 Vehicle 2 Vehicle Communication Utilizations (Application)

A wide assortment of utilizations is supported by V2V communication. From a one-hop message transfer to multi-hop transfers over significant distances as shown in Fig. 9.11.

A few utilizations of V2V communication are quickly portrayed beneath:

- **Electronic brake lights**: Electronic brake lights will enable drivers to react to the braking of vehicles, even when the vehicle is hidden by some other vehicles.
- **Platooning**: This allows vehicles to copy the driving conditions of the vehicles ahead of themselves. This is attained by wirelessly receiving the information about the acceleration and steering configurations.
- **Traffic Information Systems**: This provides obstacle reports to the vehicle's system, to help decide which paths to travel.

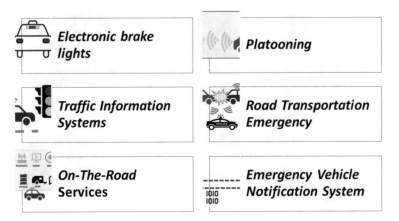

Fig. 9.11 Vehicle to vehicle communication utilizations

- *Road Transportation Emergency Services*: This provides road safety warnings and also, to speed up the rescue operations in case of emergency.
- *On-The-Road* Services: It is normal that, Future Transportation will be "Data-Driven" and "Remotely Enabled". Which implies that, while driving out and about, drivers will be informed about the services accessible close by. It is conceivable to book the film ticket while individuals are en route to films.
- *Emergency Vehicle Notification System*: If an accident occurs, then an in-vehicle "eCall" is initiated, either by the driver or automatically due to activation of the sensors. This eCall transfers the data of the vehicle and connects a voice call to the nearest emergency center. A trained eCall operator is present at the center to entertain the issues of the person in need.

The data transferred has information about the time of incident, precise location and information about vehicle identification. The use of S-band Satellite communication is expected to be employed for road safety purposes. This will provide greater coverage of the eCall service.

9.7　Commercial Applications

Commercial applications will include the streaming of web services in audio or visual form or simply access to the web. Some of the commercial applications are described below as shown in Fig. 9.12.

- *Remote Vehicle Personalization*: It can be helpful in downloading or uploading of the personalized vehicle settings or diagnostics from or to the infrastructure, respectively.

Fig. 9.12 Commercial
application of V2V

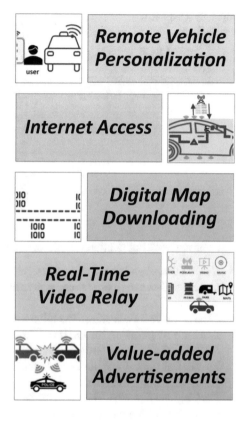

- *Internet Access*: Vehicles can be offered access to the Internet with the assistance
  of Roadside Units (RSUs) by permitting the RSUs to function as router.
- *Digital Map Downloading*: If the user decides to travel to a new place, then maps
  can be automatically downloaded, for the purpose of travel guidance.
- *Real-Time Video Relay*: Access to on demand video/movie screening can be made
  available with the help of this technology.
- *Value-added Advertisements*: It is for the most part for the comfort of service
  providers. It will empower them to broadcast advertisements, legitimately to the
  vehicles, which in the communication range. This will likewise permit the drivers
  to think about the services accessible to them, close by.

9.8   Convenience Applications

We will mostly see the ways to enhance traffic efficiency and overall traffic management methods. This will improve the vehicle driving standard, tons of benefits to the
drivers as shown in Fig. 9.13.

Fig. 9.13 Convenience
applications

The convenience applications are described below:

- *Route Diversion*: In case of road congestions, then other vehicles will be communicated to divert their directions, which initially were moving toward the congested way.
- *Electronic Toll Collection*: The toll can be gathered by the Electronic Toll Collection stall by perusing the OBUs of the vehicles. OBU is an abbreviation of On-Board-Units, it works through GPS and the technograph as a backup to decide the location of the vehicle, and GSM is utilized to approve the payment of the cost. It will be helpful to both the drivers and toll operators.
- *Parking Availability*: Vehicles can be communicated, to notify them about the availability of the parking slots in a certain area.
- *Active Prediction*: Upcoming terrain of the road can forecast to optimize the fuel consumption and also to assist the driver. So that the speed of the vehicle can be adjusted accordingly.

9.9 Summary

This presentation has introduced vehicular communication that incorporates vehicles to communicate with one another, security measures during the communication, upgrades of traffic management, immediate crisis services and implementation of assistance driver help, giving guidance and route optimization of vehicles and furthermore give the review of Intelligent autonomous vehicle works and it's safety inside an improved environment. It has additionally considered the utilizations of vehicle communication it comprises of electronic brake lights, platooning, traffic information framework, street transportation crisis service, on-the-road services. In business applications, it has given the overview of remote vehicles diversion, web access inside vehicles, computerized map downloading, real video record and flash, value-added promotions, and convenience applications has discussed remote route selection, electronic toll collection, smart parking analysis, and availability and demonstrated the vehicle-to-vehicle communication protocol standards.

Acknowledgements This research was funded by Woosong University Academic Research in 2021.

References

1. M. Singh, S.-G. Lee, W.K. Tan, J.H. Lam, Impact of wireless mesh networks in real-time test-bed setup. Adv. Inf. Sci. Serv. Sci. (AISS) Scopus J. **3**(9), 25–33 (2011). ISSN No: 2233-9345
2. I.A. Abbasi, K.A. Shahid, A review of vehicle-to-vehicle communication protocols for VANETs in the urban environment. Future Internet **10**, 14 (2018)
3. C.C. Sobin, V. Raychoudhury, G. Marfia, A. Singla, A survey of routing and data dissemination in delay tolerant networks. J. Netw. Comput. Appl. **67**, 128–146 (2016)
4. S. Zeadally, J. Guerrero, J. Contreras, A tutorial survey on vehicle-to-vehicle communications. Telecommun. Syst. **73**, 469–489 (2020)
5. R. Beraldi, A. Mtibaa, A.N. Mian, CICO: A credit-based incentive mechanism for cooperative fog computing paradigms, in *Proceedings of the IEEE Global Communications Conference (GLOBECOM)* (Abu Dhabi, UAE, 2018), pp. 9–13
6. A. Agrahari, D. Singh, Smart city transportation technologies: automatic no-helmet penalizing system, in *Blockchain Technology for Smart City* (Springer Nature, Singapore, 2020).https://doi.org/10.1007/978-981-15-2205-5_6
7. https://www.epa.gov/ghgemissions/sources-greenhouse-gas-emissions

Chapter 10
Vehicle Mobile Data Analysis for Driving Safety and Security

Madhusudan Singh ⓘ

Abstract When we are reading about data analysis research, we often hear the terms machine learning. In this article has mentioned that machine learning is shaped by unsupervised learning, supervised learning and reinforcement learning. The vehicle data generated have known parameters. Based on this fact the machine learning model that should be used would be supervised learning because we have the vehicle data as well as corresponding labels for them. This article has provided the overview of about vehicle data collection and analysis on real-time. It has mentioned the machine learning approach in vehicle data analysis.

Keywords Vehicle data · Machine learning · Mobile data analysis · Supervised learning

10.1 Introduction

Understanding the vehicle data analysis process is very important. We can see the graphical representation in a step-by-step process in Fig. 10.1. In the first step, we collect or gather) the mobility data from the vehicle through the OBD device. Then we prepare the data by pushing it through a data preprocessing cycle which cleans and formats it in a sensible format. The third step involves choosing an appropriate machine learning model [1]. This choice depends on the type of expected output. Unlike the recent past, today a lot of algorithms exist to choose from. Some examples are linear regression, support vector machine, neural networks, etc. After choosing a suitable model, in the fourth step we train the model on the analyzed data. The result It will generate an output value which we must evaluate in the fifth step in terms of accuracy. If it provides value that is satisfactorily accurate, then it is desirable. If this is not the case, we must proceed to step six where we perform hyperparameter tuning. We must choose a set of optimal hyperparameters for the learning algorithm in order

M. Singh (✉)
School of Technology Studies, Endicott College of International Studies, Woosong University, Daejeon, Republic of Korea
e-mail: msingh@wsu.ac.kr

M. Singh, *Information Security of Intelligent Vehicles Communication*,
Studies in Computational Intelligence 978,
https://doi.org/10.1007/978-981-16-2217-5_10

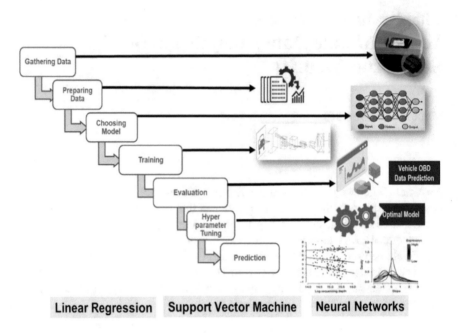

Linear Regression   Support Vector Machine   Neural Networks

Fig. 10.1 The vehicle data analysis process

to obtain an optimal model. When this is accomplished, we move to the last step, step seven where we can finally perform a prediction on a the new never-seen-before data generated from real-world data [2]. These very important seven steps consist comprise the vehicle data analysis process as shown in Fig. 10.1.

This article is organized as follows. Section 10.2 provides an overview of machine learning in vehicles data analysis. Section 10.3 discuss research on vehicular data economy and safety. Section 10.4 presents performance analysis of safety economy driving index. Section 10.4 discuss the performance analysis. Section 10.5 conclude an overview of machine learning in vehicle data in economy and safety and future research directions.

10.2 Machine Learning in Vehicle Data Analysis

Further, any problem in supervised learning can be divided into types such as regression and classification. Regression problems involve finding or predicting a future value based on the past recorded values. Regression models deal with the values that are continuous in nature. Classification problems require classifying the values in the dataset into one of the possible N groups [3]. Classification models attempt at classifying a new value among the existing determined classes from the model training.as represented in Fig. 10.2.

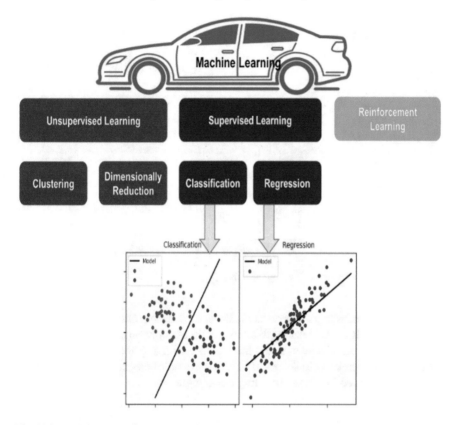

Fig. 10.2   Vehicle data analysis with machine learning via supervised learning

10.2.1   Machine Learning Modeling

The machine learning modeling phase is essential in light of the fact that the model learns from the parsed information and utilizations what it has figured out how to find patterns of interest. This is practiced by utilizing the neural network or artificial neural system which is one lot of algorithms utilized in machine learning for demonstrating or modeling the information utilizing charts of neurons. The convolutional neural network is a subfield of machine learning worried about an algorithm called artificial neural networks [4]. This is a graphic portrayal of a convolutional neural system where we start from parsed information on the extreme right and end up with patterns of enthusiasm on the extreme left. In view of the forecasts offered by the convolutional neural network we will process the real-time vehicle information as presents in Fig. 10.3.

In the graph, you can see the linear regression model. The motivation behind linear regression is to "foresee" the estimation of the reliant variable dependent on the estimations of at least one or more variables. The linear regression model

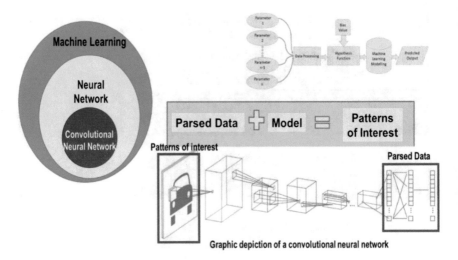

Graphic depiction of a convolutional neural network

Fig. 10.3 Example of machine learning modeling

embraced here, measures the level of the linear relationship between the two factors used for data analysis. In order for this to work, the assumption is that the relationship between data input and data output is linear. The resulting graph (Fig. 10.4) between data and linear regression looks exactly like our example here, the blue dots represent the data, and the red line is the resulting linear regression.

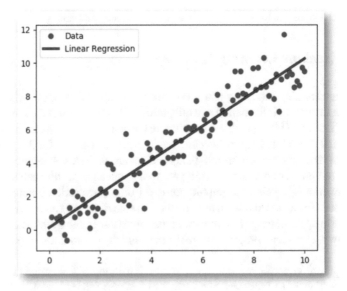

Fig. 10.4 Generated liner regression mode

Equation of Linear Regression Model:

$$\mathbf{Y_i} = \boldsymbol{\beta}_0 + \boldsymbol{\beta}_1 \mathbf{X_i} + \boldsymbol{\varepsilon_i} \tag{10.1}$$

where $\mathbf{Y_i}$ = represents the output value, $\boldsymbol{\beta_i}$ = represents the different weights for each variable, Xi = represents the variables for the model.

Let us take close look at our condition: y is the dependent variable, beta 0 is the bias value in the regression model, x is the independent variable, beta 1 is the heaviness of the independent worth x in its contribution for anticipating y, epsilon 1 is the random error component [5], it exists in light of the fact that our linear regression model attempts to imitate a real-world scenario where a perfect equation is difficult to obtain. The linear regression model has limitations but there are some improvements we can make to improve accuracy.

For several inelastic reasons linear regression models have limitations. These limitations become apparent when employed to scale over 100,000 data points. The vehicular environment is so complex and rich with data that this limit is easily surpassed. Since data size in this environment is tremendously large and there are complex relationships among the parameters themselves, the model becomes less accurate when the data size increases.

One approach to improve model accuracy is to use another sort of machine learning model known as the Neural Networks Model. Neural systems are a lot of algorithms, displayed freely after the human cerebrum, that is intended to perceive designs even in an exceptionally enormous measure of data, thusly, these systems are truly adept at finding and grouping complex connections among the related parameters. In the vehicular environment, neural networks utilize a set of algorithms that can be trained with the vehicle data descriptors for model recognition. More specifically we use the Dense Neural Network Model. We can utilize the large amount of data to train on this model to improve the accuracy of prediction. This is especially true regarding image data.

10.2.2   Vehicular Data Analysis Process

Diving deeper in the data analysis process, this is a model specifically designed for the vehicular environment [6]. It can detect and classify real-time vehicle data for the purpose of predicting outputs regarding safety and economy. Let us examine how it works as shown in Fig. 10.5: To begin with, all the parameters are inputted for data processing. In light of our theory, we have incorporated a function to help the machine learning modeling to foresee an output. It must be comprehended that in any model there is a bias value; bias is the difference between the normal expectation of our model and the right worth which we are attempting to anticipate. We attempt to restrict the bias value in this model by good data exploration, preprocessing, feature engineering, model training, and model evaluation.

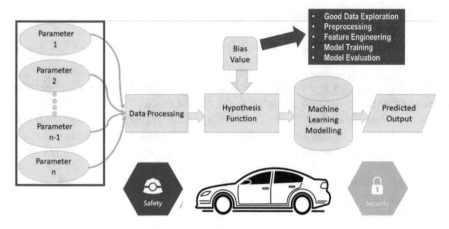

Fig. 10.5 Example of vehicle data analysis process

The mobility data analysis process is far from simple. There are five steps to consider: data collection, data processing, feature engineering, model training, and driving skill value regression. To start with Firstly, data is gathered from the vehicle utilizing the OBD port. Second, the data is prepared by the normalization model. Third, feature engineering is designed upon the correlation values. Fourth, the model is prepared for the linear regression model [7]. Fifth, the driving skill value regression predicts values on various sorts of driving skill that are obtained as depicted in Fig. 10.6.

As stated earlier, in any model there is a bias value and hypothetical function generation methods can limit that bias and make our results more accurate. In order to limit bias further, two methods can be followed. First, we assigned different vehicles for each variable. For example: the variable of fuel efficiency was assigned to 6000 different kinds of vehicles. Hypothetical Function Generation methods to Limit Bias has explained in Eqs. 10.2 and 10.3.

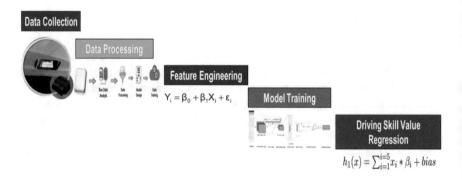

Fig. 10.6 Mobile data process

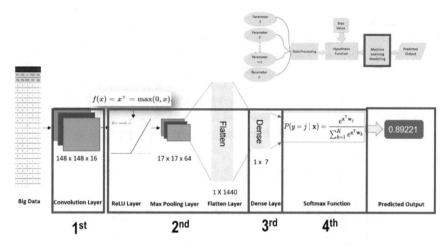

Fig. 10.7 Convolutional neural network model for vehicle

$$y = \beta_0 + \beta_1 * x_1 + \beta_2 * x_2 + \beta_3 * x_3 + \beta_4 * x_4 + \beta_5 + x_5 \qquad (10.2)$$

$$y_1 = \beta_0 + \beta_1 * x_1 + \beta_2 * x_2 + \beta_3 * x_3 + \beta_4 * x_4 + \beta_5 + x_5 \qquad (10.3)$$

Second, we chose the Convolutional Neural Network Model because it utilizes a set of algorithms that can be trained with the vehicle data descriptors for model recognition. Each hypothesis is based on the Linear Regression Model. Looking at the equation, y_i represents the output value, Beta i represents the different weights for each variable and X_i represents the variables for the model.

Using the Convolutional Neural Network Model for vehicle video data makes a lot of sense [8]. As explained earlier, we utilize the Convolutional Neural Network Model for object recognition purposes and video falls under this category. It helps with the classification of big data and helps train sets of algorithms with vehicular data to create model recognition. This graph shows how the convolution neural network model consists of different layers. First, big data goes to the convolution layer. Second, the data goes to three fully connected layers: the rectified linear unit layer, the max pooling layer and the flatten layer. Third, the data goes to the dense layer which is the final classifying layer. Fourth, the data is classified, and we use the Softmax Function to help predict the output results as presented in Fig. 10.7.

10.3 Research on Vehicular Data Economy and Safety

Even the best model in the world is useless if you have bad data. In our research is based on, the OBD-II port data and is supplied and communicated in real time. The graphic on the right shows us the information from one vehicle and only those data points regarding speed RPM and throttle voltage from the engine. As you can see, we have over 1200 data points from which we can derive conclusions regarding safety and economy. This is accomplished after several levels of pre-processing data to make it cleaner and convenient to read and analyze for further action. This graph was derived after several levels of data preprocessing to make it cleaner and convenient to read and understand. As represented in Fig. 10.8.

In Table 10.1, we can see safety can affect the directly on economy and its improved the efficiency of vehicle with help of vehicle data analysis model.

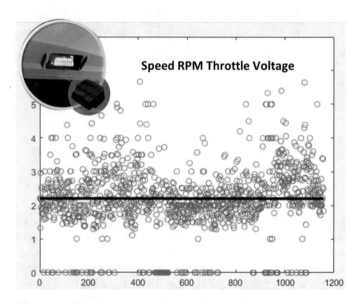

Fig. 10.8 OBD-II port for vehicle data collection-real time

Table 10.1 The effect of safety on economy for vehicles

Safety	Economy
Average and maximum speed Average braking and accelerating behavior	Fuel efficiency Driving conducive to unnecessary engine wear
Number of emergency accelerations	Vehicle travelling distance and navigation choices
Erratic maneuvering	Green zone driving time value
Long driving time exceeding driver endurance	–
Red zone driving time value	–

10.3.1   Using Linear Regression Model for Vehicle OBD Data Processing

Economy driving indexes give an estimate of the state of a vehicle and to what extent it will run. Following one month of experimentation on a fleet of roughly 100 participating vehicles by the Mission to Vision Company in South Korea, the information gathered by means of the OBD-II scanner introduced in every one of the test vehicles was analyzed. This was done to train the algorithm. We recorded approximately 6000 data points that were divided by an 80–20 ratio, where 80% were used for training and 20% were used for testing. Afterwards, using a correlation matrix, we were able to derive useful information regarding our purpose to derive an Economy Driving Index. Figure 10.9 presented the process of linear regression model for vehicle OBD data processing.

In Table 10.2, we show what parameters we studied to calculate an Economy Driving Index based on our initial hypothesis. The five parameters used were: Fifth Throttle RPM time, Urgent Acceleration Number, Urgent Deceleration Number, and Fourth Throttle RPM time. Looking at the results listed under the correlation value column, we were able to understand the relationships between the five parameters.

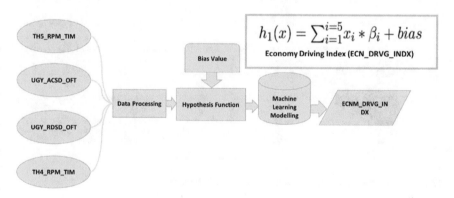

$$h_1(x) = \sum_{i=1}^{i=5} x_i * \beta_i + bias$$

Economy Driving Index (ECN_DRVG_INDX)

Fig. 10.9   Processing real time vehicle data modelling

Table 10.2   Parameter of economic driving index (ECN_DRVG_INDX)

ECNM_DRVG_INDX	Correlation matrix	
	Features	Correlation value
TH5_RPM_TIM	Fifth throttle RPM time	−0.567331
UGY_ACSD_OFT	Urgent acceleration number	−0.615989
UGY_RDSD_OFT	Urgent deceleration number	−0.621209
TH4_RPM_TIM	Fourth throttle RPM time	−0.563859

10.3.2 Performance Analysis of the Economy Driving Index

In Fig. 10.10 Has introduced the exhibition of the economy driving index. Where the blue dots are our data points, and the orange line is the prediction line. We found that a large chunk of the data was clogged in the lower-left piece of the graph recommending an inverse logarithmic growth of the pattern dependent on the training data. This indicated a positive development of the Economic Driving Index dependent on the speculation value.

A similar data analysis process pipeline was developed for calculating the safety driving index as well. This gives us an overview of how cautiously the driver is driving the vehicle. The derived Safety Driving Index is shown on the top right. As we can see in Fig. 10.11.

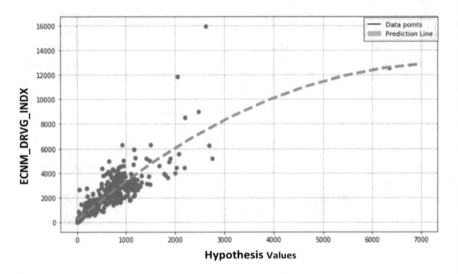

Fig. 10.10 Performance analysis of the economy driving index

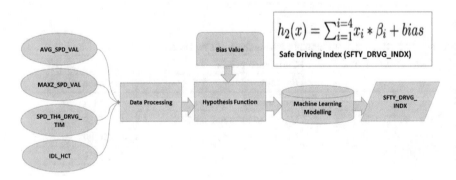

Fig. 10.11 Safety driving index using linear regression model for vehicle OBD data processing

Table 10.3 Processing real time vehicle data modelling

SFTY_DRVG_INDX	Correlation matrix	
	Features	Correlation value
AVG_SPD_VAL	Average speed value	0.951148
MAX_SPD_VAL	Maximum speed value	0.850964
SPD_TH4_DRVG_TIM_VAL	Fourth throttle driving time value	0.676948
IDL_HCT	Idle time value	−0.296508

In Table 10.3, we show what parameters we studied to figure the Safety Driving Index dependent on our underlying theory. The four parameters utilized were: Average Speed Value, Maximum Speed Value, Fourth Throttle Driving Time Value, and Idle Time Value. Taking a look at the outcomes, recorded under the correlation value section, we had the option to comprehend the connections between the five parameters.

10.3.3 Performance Analysis of the Safety Driving Index

In Fig. 10.12 shows the performance analysis of the safety driving index. Where the blue spots are data points, and the orange line is the forecast line. We watched a slower beginning development in the pattern yet then it quickly gets force in the wake of intersection a specific threshold value. After this threshold value, the pattern

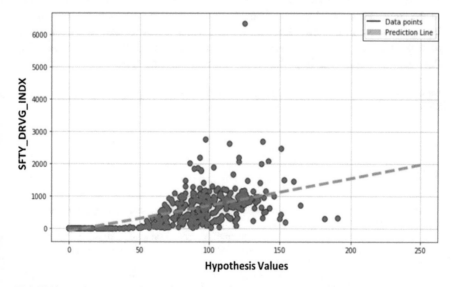

Fig. 10.12 Performance analysis of the safety driving index

moves in a logarithmic manner upwards towards the positive y-axis. This indicated a quick positive development of the Safety Index Value dependent on the hypothesis values.

10.4 Discussion

Using these data parameters and the model output values, we can infer that to ensure a safe driving scenario, we must not press the brakes hard, engage in hard acceleration, speed excessively, or make sudden lane changes, etc. Similarly, we can infer that right-time supply replacement and fast navigation must be adopted to ensure a smart driving scenario. There can be many more possible inferences once the data is cleaned and processed in a mathematical fashion using machine learning. The real-time vehicle data performance analysis predicts that safety and smart driving results.

10.5 Summary

In this article we have discussed the safety for intelligent autonomous vehicles, collection and understanding of available real-time vehicular data, understanding the analysis process and interpreting all available data, and also guide future research findings regarding vehicular economy and safety. A data driven predictive approach certainly defines the upcoming approach to decide the next action to be taken on real time basis. Every data and information collected from sensors and routers placed on the roadsides as well as data from other vehicles are computed on data driven approach based on machine learning model, which helps to take best decision as possible. Here in this article, we have learned and viewed how machine learning model approach helps to improve safety and smart driving results.

Acknowledgements This research was funded by Woosong University Academic Research in 2021.

References

1. R.K. Ganti, F. Ye, H. Lei, Mobile crowdsensing: current state and future challenges. IEEE Commun. Mag. **49**(11), 32–39 (2011)
2. P. Mohan, V.N. Padmanabhan, R. Ramjee, Nericell: rich monitoring of road and traffic conditions using mobile smartphones, in *Proceedings of 6th ACM Conference on Embedded Networking Sensor Systems*, pp. 323–336 (2008)
3. K.L. Clarkson, Algorithms for Closest-Point Problems (Computational Geometry), Ph.D. Dissertation. Stanford University, Palo Alto, CA. UMI Order Number: AAT 8506171 (1985)

4. N. Fan, Z. Duan, G. Zhu, A data dissemination mechanism based on evaluating behavior for vehicular delay-tolerant networks. Int. J. Distrib. Sens. Netw. **15**, 1550147719865509 (2019)
5. Y. Zhan, Y. Xia, J. Zhang, Y. Wang, Incentive mechanism design in mobile opportunistic data collection with time sensitivity. IEEE Internet Things J. **5**, 246–256 (2017)
6. D. Peng, F. Wu, G. Chen, Data quality guided incentive mechanism design for crowdsensing. IEEE Trans. Mob. Comput. **17**, 307–319 (2017)
7. A. Schneider, G. Hommel, M. Blettner, Linear regression analysis: part 14 of a series on evaluation of scientific publications. Deutsches Ärzteblatt Int. **107**(44), 776–782 (2010)
8. P. Yadav, S. Jung, D. Singh, Machine learning-based real-time vehicle data analysis for safe driving modeling, in *The 34th ACM/SIGAPP Symposium on Applied Computing (SAC)* (Limassol, Cyprus, April 8–12, 2019), pp. 1355–1358

Chapter 11
Vehicular Data Analytics and Research Findings on Security, Economy and Safety

Madhusudan Singh[ID]

Abstract This presentation has discussed Cyber security and safety for autonomous vehicles with the understanding of real time vehicular data. It has also given an understanding the analysis process and interpreting the vehicular data and finally conclude the course with the explanation of automotive cybersecurity research findings in economy and safety areas. This presentation has presented the vehicular data analytics and research findings on security, economy and safety. It provides cybersecurity and safety for intelligent autonomous vehicles, collection and understanding of available real-time vehicular data, understanding of analysis process and show process of data interpretation and finally it's discussed research findings to be drawing and conclude the impact of vehicular data in economy and safety.

11.1 Introduction: The Fundamentals of Automotive Cybersecurity

The basics of automotive cybersecurity focusing on safety and security. These are defined depending on the source of the threat. The formal definition of safety is the condition of being protected from the harm caused by non-intentional failure. It describes a situation when acquired values are harmed by accidental flaws and mistakes. It occurs when accidental flaws and mistakes occur such as technical errors, network failures, environmental disasters etc. It protects against potential or actual harm to acquired values. In the vehicular network, harm can occur at any time due to environmental causes or technical errors. Security, on the other hand, is the condition of being protected from harm caused by intentional human action or behaviors; it protects against potential or actual harm to acquired values [1].

If we take a closer look at similarities and differences between safety and security, we can see in Fig. 11.1 that, in essence, both concepts are about potential or actual

M. Singh (✉)
School of Technology Studies, Endicott College of International Studies, Woosong University, Daejeon, Republic of Korea
e-mail: msingh@wsu.ac.kr

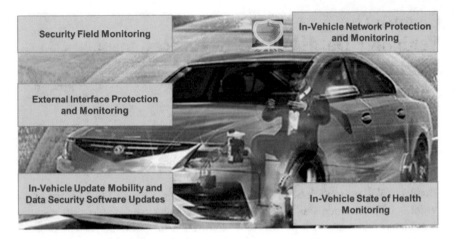

Fig. 11.1 Vehicle safety and security

harm to acquired values. When we look at safety and security needs regarding intelligent and autonomous vehicles, we introduce the concepts of protection and performance. Under safety [2], we examine knowledge and skills while under security we include roles and norms.

It is good to remember that safety and security are quality attributes that are part of an overall systems engineering process. This process has been standardized long before being implemented in the automotive industry thanks to ISO standards. Security ISO standards are established by first defining the security goals, then designing the functional security concepts, which is followed by designing technical security concepts, and in the last, defining software and hardware components to create a secure system.

11.2 Vehicle Safety and Security Relationship

The ultimate goal is to harmonize the security and safety aspects of automotive technology. Together, they form a wall which can provide security field monitoring, external interface protection and monitoring, in-vehicle update mobility and data security software updates, in-vehicle network protection and monitoring, and in-vehicle state of health monitoring. It's shown in Fig. 11.1.

11.2.1 Automotive Safety Begins with Security and Reliability

Without security and reliability, there is no automotive security. We will concentrate on device dependability, functional security, and on-time working safety. Device

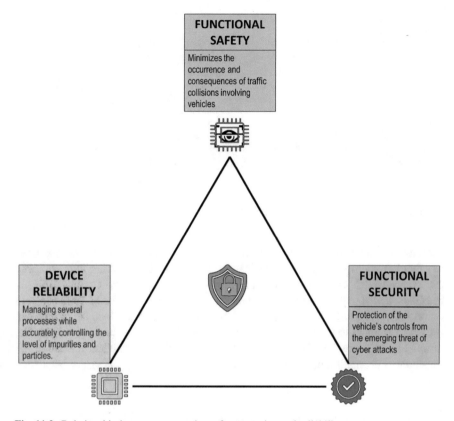

Fig. 11.2 Relationship between automotive safety, security, and reliability

dependability is important to deal with a few procedures while precisely controlling the degree of impurities and particles from entering the hardware before it can cause inside harm. It represents in Fig. 11.2. This is generally ensured by using ISO-compliant products. Functional Security is the protection of the vehicle's controls from the emerging threat of cyber-attacks; these threats are highly dangerous in the automated driving scenario [3] where the driver has little to no control over the vehicle. Functional Safety is provided by the automotive manufacturer to minimize the occurrence and consequences of traffic collisions involving vehicles.

11.2.2 In-Vehicle Security Threats

In-vehicle security threats are a big component of the cyber security puzzle. These threats deal with the possible channels inside a vehicle that can be targeted for the purpose of compromising the security of the entire vehicle. There are primarily three potential methods to breach vehicular security: telematics hacking, smart phone

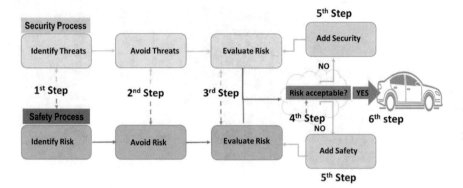

Fig. 11.3 Automotive security versus safety process

hacking, and stealing and hacking into the vehicle's OBD port. In the diagram, we see that the way to break into a vehicle through telematics hacking is to link into the external NIC which can ultimately provide a passage to the Control Unit via the Gateway [4]. By hacking into the smartphone of the driver, the hackers can gain control of the internet-connected components such as ADAS, TMS, etc. which can provide a channel to the Control Unit through the Gateway by stealing the vehicle, or more specifically the OBD port, installed in the vehicle, the hacker that can reveal invaluable information about the driver and the vehicle itself.

11.2.3 Outside Vehicle Security Threats

Figure 11.3 has shown the potential in vehicle security threats, this section as discussed, the threats on the exist outside of the vehicle. All the electronic systems that are part of the vehicular environment, these are: the traffic management system, the roadside equipment (such as traffic lights), the nearby environment, software delivery/updates, the service cloud and the user's mobile device. All six of these parts of the vehicular environment can threaten the security of the vehicle. The threats can affect all features of the system from traffic efficiency and safety, to fees and charges, vehicle interaction, and even the on-board infotainment system [5].

11.2.4 Automotive Security vs. Safety Process

The security and safety validation cycle components one by one starting from step one of the security process. The first step is to identify the threats to the vehicle. In the second step we try to avoid security threats. In the third step, we evaluate risk. Once the risk is evaluated, we determine whether the risk is under an acceptable limit or not. If it is not, then we deploy more security and safety features in the vehicular system [6]. We proceed only when we determine that the risk is acceptable. The safety process is very similar to the security process as we can see in Fig. 11.3. Since automotive security is ever evolving, so should the safety process that guards it. To achieve this, there is a very stratified process.

11.3 Vehicular Cloud Management System

The typical connected vehicle system consists of several technologies such as internet, cellular network, radio technology, Bluetooth and many more. These technologies serve as the communication channel for data transmission in the cloud system. That is where the data collected from the vehicle goes for processing and analysis. In this graphic, we can see an overview of the vehicular cloud management system which is incorporating everything from audio-video services to IT systems [7]. This cloud system serves as the all-inclusive hub for all the management and processing activities conducted on the network.

The all-inclusive possible protection from the cyber-attacks, integrate consumer devices, provide car-to-car and car-to-infrastructure communication through a secure channel [8]. It also has an on-board safety mechanism to keep the vehicle safe from the inside. This possible only by achieving what is called harmony among safety ad security. It is depends greatly upon the level of trust between security and safety.

11.4 Vehicle Mobility Data Processing Methods

With regards to vehicle mobility information handling, things get precarious. There is an absence of uniform tasks and a standard portrayal of the vehicle's sensor collected data. Right up 'til the present time, there is no cloud classification for vehicle asset reallocation and asset sharing of data because of absence of interoperability. In Fig. 11.4 above we depict this problem trying to show that we have heterogeneous data, a high number of machines in the vehicle and data complexity. This is the reality but what do most people think about when autonomous vehicles are mentioned?

In below graph shows how perceptions regarding autonomous vehicles are changing. Here we can see that the public's perception has changed positively in seven major industrial countries of the world. All this within only one year. Based on this big leap forward shows that with people's trust increasing that would mean higher market expectations for autonomous vehicles in the near future as presents in Fig. 11.5.

The vehicle mobility data market size is growing and in later years we have seen continued acceleration of investments in all relevant technologies [9]. Currently, the shared mobility and the data market is worth $30 billion. By 2030, it is expected

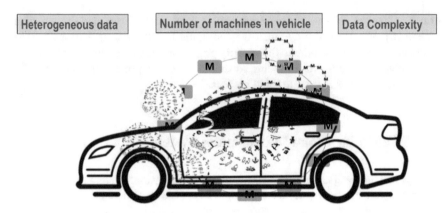

Fig. 11.4 Vehicle mobility data processing

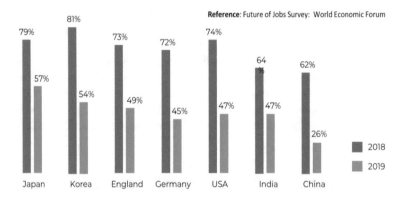

Fig. 11.5 Graph of perceptions regarding autonomous vehicles are changing

to grow over to over $2 trillion according to the McKinsey & Company's analysis report.

11.4.1 Overall Volume of Connected Car Data Transfer

While talking about the general volume of connected vehicle information transfer, we are anticipating gigantic development. This development will happen because there will be a tremendous increment in the volume of information move per-vehicle and it is evaluated that the absolute number of connected vehicles on the planet will increment at a fast rate with time represents in Fig. 11.6.

Increase in the volume of
data transfer per-vehicle

Increase in total number of
connected vehicles

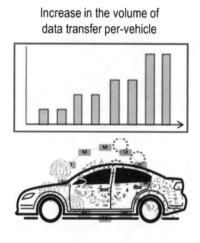
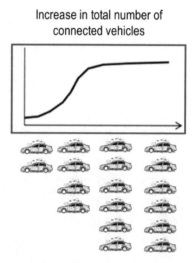

Fig. 11.6   Volume of connected car data transfer

11.4.2   Connected Vehicles Generated Mobility Data Collection Methods

It is clear that connected vehicles generate a lot of mobility data but what kind of data are we looking for and what collection methods do we employ? We are looking for driving history data, vehicle status data, and driving behavior data. Driving history data is all the data stored by the vehicle during operation and the location data for when the vehicle is in motion. The vehicle status data is the kind of data that includes current status of the battery, engine temperature, fuel quantity, tire pressure, etc. Last, "driving habit data" as it is known includes actual data such as that of rapid acceleration, sudden stopping, etc. Driving history data such as location and speed data can be collected using connected mobile devices with the help of GPS. Vehicle Status data can be collected using the OBD-II device plugged into the vehicle. Driving behavior data can be collected using vehicle to everything communication as presented in vehicle mobility data collection.

If we look a bit deeper into what are the sources of all these mobility data points, we can observe that modern GPS probes and location-based services are generating hundreds of MB of data per month. Even the simplest dynamic map generation and real-time user tracking require 100 s of MB. The automotive electronic control unit that controls at least one of the electrical systems or subsystems in a vehicle likewise creates a few several MB data for every month. Last, even the peripheral detecting modules introduced in autonomous vehicles are producing comparative measures of data every month. Figure 11.7 has introduced the sources of vehicular mobility data.

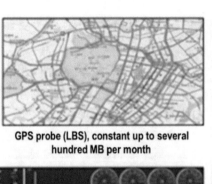

GPS probe (LBS), constant up to several
hundred MB per month

Dynamic map generation, as needed up to
several hundred MB per month

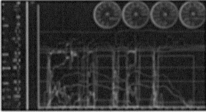

ECU generates data, as needed up to several
hundred MB per month

Peripheral sensing data, as needed up to
several hundred MB per month

Fig. 11.7 Sources of vehicular mobility data

11.4.3 Connected Vehicle Sensors and Data Collection

Connected vehicles are a constellation of computer chips and mechanical, electrical, electronic sensors that are managed by information technology. One main purpose is data collection in ever increasing volumes since even current, non-connected vehicles process up to 25 gigabytes of data an hour. One can only imagine how much computing power a connected vehicle requires to manage vision, guidance and the mapping technology in order to process this ever-growing volume. In the 1970s, vehicles were 100% hardware while the projections for 2025 forecast that almost half will be software and new applications.

11.5 Vehicle Data Processing

Vehicle data collecting and processing is of paramount importance. The data is meaningless if nothing happens after it is collected. In its raw form, data is of no use. The decentralized data collected must be uploaded to the cloud where raw data analysis, data processing, model design and data training takes place. We do all of this to get to meaningful information as presented in Fig. 11.8

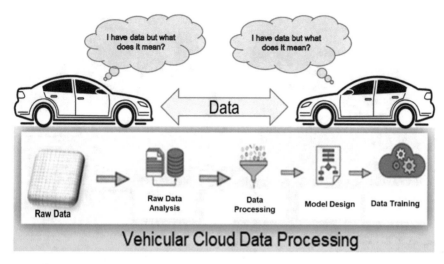

Fig. 11.8 Vehicular data processing

After collecting various types of data from the vehicle, it is processed at a high level before to extract meaningful information that can be used to improve the vehicle efficiency, driving behavior, vehicle safety, etc.

In the above table on the left side we can see, what the collectible data looks like listed with names and descriptions. Respectively, on the right. The table shows again the data point name with a description of the actual information we obtain after the processing stage. Next, we will analyze the whole workflow from data collection, to information generation, to the final upload to the cloud.

11.5.1   Vehicle Cloud Data Processing

As we know that vehicle cloud data processing can be very tricky because we are talking about a lot of data that is both heterogeneous and complex. Let us look at the whole workflow from data ingestion to information generation in the cloud. After the collection, in the data ingestion platform the raw data along with the driver and vehicle meta data is ingested in a data lake. From there it is further processed by applying data cleaning, data exploration and feature engineering techniques. Now the data is ready to be put into a data normalization model or a machine learning model. These models can generate information and make relevant predictions for business functions such as insurance, car sharing, logistics, etc. The final information is displayed on a dashboard or a smart phone application. This information process generation runs in a cycle with new data continuously pouring in. Since the process runs over and over again, it leads an ever "smarter" or more refined model and thus provides more and more trustworthy information as represents in Fig. 11.9.

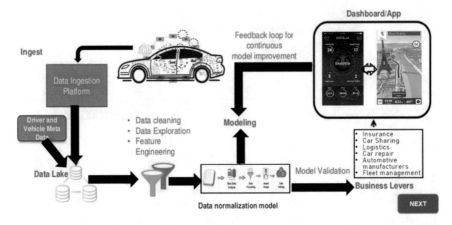

Fig. 11.9 Workflow of vehicle data processing

11.5.2 Vehicle Mobility Data Services

The present and future mobility data services. In the current environment, vehicle mobility data services are already is use in the areas of location-based solutions, determining a vehicle life cycle, in corporate vehicle management, and in services such as insurance, etc. However, in the future we expect that the mobility data will also be used for the vehicle sharing management system that it will provide and improvise autonomous driving solutions, that manufacturers will be able to purchase data from the vehicle owner, that maintenance and accident history management will be available, that non-identified data utilization solutions will be implemented, and finally that biometric data collection will also be employed and utilized. With all these new and exciting possible future applications, the value and availability of the vehicle mobility data will go up exponentially.

The current situation regarding connected services offered by the Toyota Car manufacturing company. This particular car manufacturer offers emergency notification service, theft tracking, automatic map data updates, and operator assistance as shown in Fig. 11.10. This is possible mainly because of the OBD-II device mounted on these cars that provide real-time data. The manufacturer's smart center provides various additional services such as traffic information and "look ahead" information services that stem from the big data vehicular database.

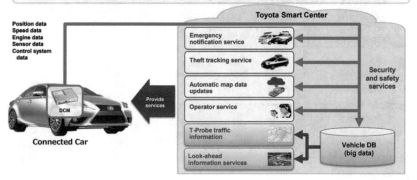

Fig. 11.10   Example Toyota vehicle connected services

11.5.3   Services Using Mobility Data

It is obvious that different kinds of data collected can be categorized based on the types and the services they can contribute to. Here under data category, we list the kinds of data collected. Such data as those regarding road condition, traffic volume, etc. are used today for real-map service for navigation but can also be used in the near future to provide a real-time road situation report which the driver can use in determining an efficient and safe route to their destination. It can also contribute to a 3-step conditional autonomous driving which means that the vehicle can take over all driving functions under certain circumstances. Data that describes vehicle condition is collected today by the OBD-II and can make emergency phone calls if there is an accident. In the future, it can be used in predicting vehicle functionality anomalies and help the driver schedule services remotely. It can also keep a certified and accurate vehicle history stored in the cloud that could be used when the car is sold in the future to determine the resale value. Last, vehicle usage data are used today by insurance companies to offer customized contracts and by fleet management systems. In the future, this data can be used to enhance the vehicle sharing platform services, facilitate large vehicles with special cargo on one control platform and share economic logistics infrastructure which is an essential part of all logistic systems as shown in Fig. 11.11.

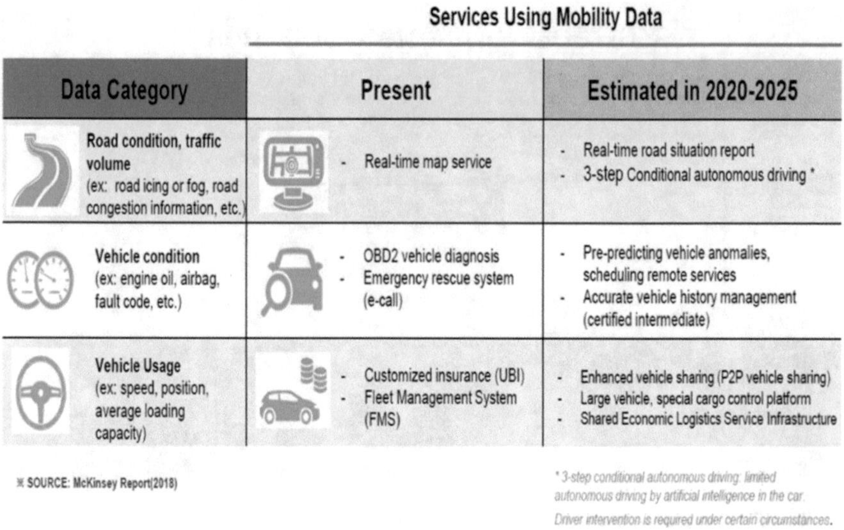

Services Using Mobility Data

Data Category	Present	Estimated in 2020-2025
Road condition, traffic volume (ex: road icing or fog, road congestion information, etc.)	- Real-time map service	- Real-time road situation report - 3-step Conditional autonomous driving *
Vehicle condition (ex: engine oil, airbag, fault code, etc.)	- OBD2 vehicle diagnosis - Emergency rescue system (e-call)	- Pre-predicting vehicle anomalies, scheduling remote services - Accurate vehicle history management (certified intermediate)
Vehicle Usage (ex: speed, position, average loading capacity)	- Customized insurance (UBI) - Fleet Management System (FMS)	- Enhanced vehicle sharing (P2P vehicle sharing) - Large vehicle, special cargo control platform - Shared Economic Logistics Service Infrastructure

※ SOURCE: McKinsey Report(2018)

* 3-step conditional autonomous driving: limited autonomous driving by artificial intelligence in the car. Driver intervention is required under certain circumstances.

Fig. 11.11 Services using mobility data

11.6 Future Data Mobility for Connected Vehicles

The future mobility data management for connected vehicles, the process can be summarized into three steps: First is data collection via the OBD-II scanner which can detect malfunction in advance. Second, from the vehicle we process all of the collected driving data and from this data we can determine the driver's driving pattern. Finally, by conducting data cross analysis, we can create new insights through cross analysis of collected and heterogeneous data. This can offer us insight in the driver's behavior as shows in Fig. 11.12.

11.7 Summary

In this chapter we gave discussed the vehicle data Safety and Security, relationship between Automotive safety, security and reliability, In vehicle Security threats, outside Vehicle security threats, automotive security versus Safety Process, vehicular cloud management system, automotive trust between safety and security vehicle mobility data processing, vehicle mobility data market size, volume of connected car data transfer, vehicle mobility data collection. Also shows the sources of vehicular mobility data, connected vehicle sensors and data collection, vehicular Data

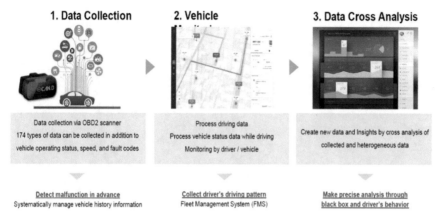

Fig. 11.12   Future data mobility for connected vehicles

Processing and workflow of vehicle data processing, services using mobility data, and discussed the future data mobility for connected vehicles.

Acknowledgements   This research was funded by Woosong University Academic Research in 2021.

References

1. R. Pal, A. Prakash, R. Tripathi, D. Singh, Analytical model for clustered vehicular ad-hoc network analysis. ICT Express (2018). https://doi.org/10.1016/j.icte.2018.01.001
2. J.E. Meseguer, C.T. Calafate, J.C. Cano, P. Manzoni, Driving styles: a smartphone application to assess driver behavior, in *2013 IEEE Symposium on Computers and Communications (ISCC)*, Split (2013), pp. 000535–000540. https://doi.org/10.1109/ISCC.2013.6755001
3. Y. Zhang, W. Lin, Y. Chin, Data-driven driving skill characterization: algorithm comparison and decision fusion. SAE Technical Paper 2009-01-1286 (2009). https://doi.org/10.4271/2009-01-1286
4. D. Singh, M. Singh, I. Singh, H.-J. Lee, Secure and reliable cloud networks for smart transportation services, in *The 17th IEEE International Conference on Advanced Communication Technology (ICACT2015)*, Phonix Park, South Korea (2015), pp. 358–362
5. Pal, N. Gupta, A. Prakash, R. Tripathi, Adaptive mobility and range-based clustering dependent MAC Protocol for vehicular Ad Hoc networks. Wireless Pers. Comm. 1–16 (2017)
6. Y. Chen, M. Fang, S. Shi, W. Guo, X. Zheng, Distributed multi-hop clustering algorithm for VANETs based on neighborhood follow. EURASIP J. Wirel. Comm. Netw. **1**, 98–109 (2015)
7. X. Yang, Y.K. Chia, S. Sun, H.F. Chong, Mobile data off-loading through a third-party WiFi access point: an operator's perspective. IEEE Trans. Wireless Commun. **13**, 5340–5351 (2014)
8. D. Singh, M. Singh, I. Singh, H.J. Lee, Secure and reliable cloud networks for smart transportation services, in *The 17th IEEE International Conference on Advanced Communication Technology (ICACT2015)*, Phonix Park, South Korea, 1–3 July (2015)
9. D. Singh, G. Tripathi, S. C. Shah, R.R. Righi, Cyber-physical surveillance system for the internet of vehicles, in *IEEE World Forum on Internet of Things WF-IoT 2018*, 5–8 February 2018, Singapore, pp. 551–556 (2018)

Chapter 12
Secure Vehicular Cloud for Vehicle Communication

Madhusudan Singh ⓘ

Abstract The quantity of car crashes far and wide are evaluated by the World Health Organization at around 1.3 million every year. All-inclusive, auto crashes are the main source of death among youthful grown-ups ages 15–29. Connected vehicles are conceptualized so that they share traffic data continuously upgrading security. In addition to safety, connected vehicles offer a range of advanced services to vehicle owners whether individuals that own one vehicle or companies that own fleets of vehicles, government transport authorities, vehicle manufacturers, and other service providers such as gas stations, service stations, parking lots, etc. These services increase the efficiency of the entire transportation system. The benefits are numerous, let us examine how we can achieve them. Internet connected vehicles play a vital role to improve efficiency and safety of road transportation systems.

12.1 What is a Connected Vehicle?

A connected vehicle is connected with other vehicles in a digital environment over the internet through systems that integrate advanced communications technologies into the transportation infrastructure and into the vehicles themselves. They are called Intelligent Transportation Systems or ITS. The idea stems from the fact that we are living in an era where it has become extremely easy to transmit information from one corner of the world to another. The connected vehicle in an ITS system is an essential part of the automotive industry which is growing exponentially.

Today, connected vehicles can "see" the wide view and deal with all safety basic capacities under specific conditions yet the driver is as yet expected to assume control over when cautioned. This is because with the present technology 100% trust in the connected vehicles can end up being deadly forever. This is bound to change in the future [1]. With the combination of sensors, cameras, a navigation system, software,

M. Singh (✉)
School of Technology Studies, Endicott College of International Studies, Woosong University, Daejeon, Republic of Korea
e-mail: msingh@wsu.ac.kr

© The Author(s), under exclusive license to Springer Nature Singapore Pte Ltd. 2021   169
M. Singh, *Information Security of Intelligent Vehicles Communication*,
Studies in Computational Intelligence 978,
https://doi.org/10.1007/978-981-16-2217-5_12

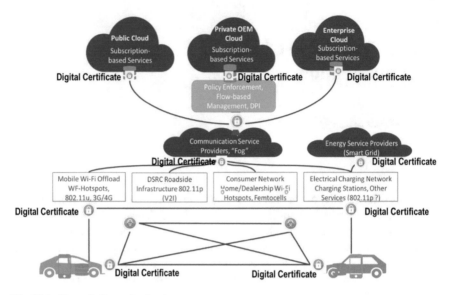

Fig. 12.1 Open platform cloud server

and established traffic rules, connected vehicles are capable of exploring and driving efficiently to their destination. It is an amazing technology where the vehicles are gradually gaining full control over driving as shows in Fig. 12.1.

12.1.1 Fundamentals of Connected Vehicles

To begin to understand connected vehicles, we must define the fundamentals that govern them. There are five fundamental areas of technology utilized by connected vehicles.

- **Edge Point** or **Internet of Vehicle(IoV)** refers to vehicle data collection, analysis, artificial intelligence, and other computing resources that are within the vehicle itself.
- **The Local Processing and Storage System** is answerable for all the inside information produced by the sensors of the vehicle.
- **Network Connectivity** involves both an internal network within the vehicle itself and an external network that is part of the ITS system.
- **Cyber Security Service Center** has the purpose to monitor and protect against potential malware infections.
- **Vehicular Cloud** is a hybrid cloud and mobile network combined in which all vehicles and infrastructure share their resources and platforms collaboratively.

12.2 The Need for an Open Platform

When we looked at the fundamental areas of technology used by connected vehicles and the amount of data that is exchanged between and within each vehicle, we can see the necessity for an open platform. An open platform must be based on an **Open Platform Cloud Server** so that all collaborators can share information seamlessly. All vehicles and infrastructure data from one country or region of the world must be able to share their vehicular data at the network level without restraint [2].

Today there is no open-access, vehicle-related data and information platform available that can do everything. The singular exception is that of navigation information. Let us now examine requirements at the vehicular level.

12.2.1 The Required Connected Vehicle System Architecture

The envisioned system architecture elements. System architecture on connected vehicles begin with the OBD Scanner and Mini Dashcam which are installed in the vehicle. The Mini Dashcam will record video outside the car while the OBD Scanner will provide the data. These two devices along with the GPS device collect necessary data about the vehicle's interior health and the surrounding conditions and through specific connectivity capabilities the data is sent to a smartphone app. This app not only has the capability of collecting but also sending the data over cloud servers for analysis. The app also acts as the main user interface for displaying the output. This data is sent over Cloud Servers which stores and analyzes it to produce desired outputs such as optimizing traffic, addressing safety and other related information. The blow Fig. 12.2 mentions data analysis going to B2B insurance, etc. Now that we understand the system of the architecture of connected vehicles [3].

12.2.2 The Vehicular Open Platform Cloud Server Paradigm

It is understood that the vehicular open platform cloud server must come with as much server space as possible and a first-class user interface (UI). This server must also include a series of specific features that most collaborators desire. These features are an Infrastructure-as-a-Service (IaaS) Cloud, a Network-as-a-Service, Compute Orchestration, client and data manager, a full and open local API and asset bookkeeping [4]. Along with these features, we should hold fast to the ordinary IT best practices, for example, security, software stacks, software updates, equipment support, and platform maintenance.

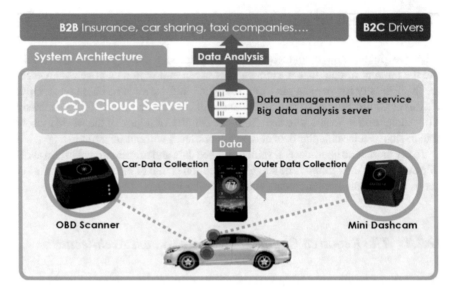

Fig. 12.2 The required connected vehicle system architecture

12.2.3 Advantages of Connected Vehicles

An overview of the advantages that connected vehicle services can offer for both vehicle owners and service providers. A vehicle owner will enjoy the following advantages: real-time driving feedback, real time cost efficiency analysis, car health remote scans, parking services, weather information, anti-theft protection.

Service providers, such as the ones listed, can expect to improve their services based on the collected data such as: more comprehensive and less expensive insurance, car sharing, logistics, car repair, automotive manufacturers and fleet management that presents in Fig. 12.3 [5].

12.2.4 The OBD II as an Automotive Vehicle Date Collection Source

The OBD II as an automotive vehicle date collection source that is existing technology that has been mandatory in every vehicle since 1996. Today, all vehicles have the OBD II port incorporated within them when manufactured. The autonomous vehicle taps into this device seamlessly and receives essential data in real time which transmits to the vehicular cloud and receives an exchange in a wide array of relevant desired services. For example: the OBD scanner and the dashboard camera can work together in collaboration with the cloud server to help the vehicle owner and a service provider discover each other in real time.

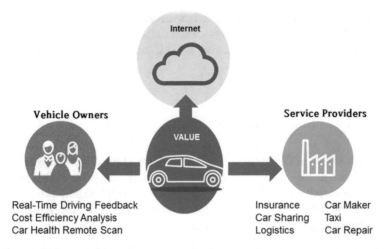

Fig. 12.3 Advantages of connected vehicles

12.3   The Automotive System Architecture

The automotive system architecture is the concept that the connected vehicle is heavily reliant on for it to function properly. Without the internet in general and an ITS system specifically designed for this purpose there will be no connected vehicles to speak of. In this system architecture as shows in Fig. 12.4, we see how connectivity

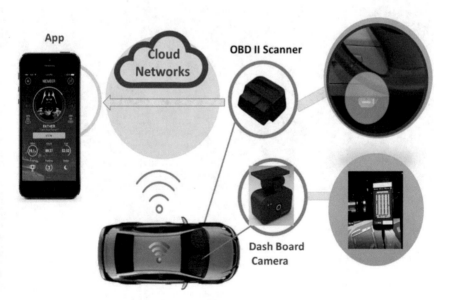

Fig. 12.4 Automotive system architecture

is established over the cellular network between automotive vehicles and the cloud server via wireless communication like 5G, Wi-Fi, etc. Interestingly enough a lot of information is supplied by the "humble" OBD II scanner that as we mentioned has been around for decades [6].

12.3.1 The Secure Vehicular Cloud Server Platform

The Secure Vehicular Cloud Server Platform is the most important component in the whole system, because all vehicle-related data is stored there, hence it is of utmost necessity to keep it secure from any third-party attacks. It basically acts as a black box, so that nothing inside is visible. The platform consists of three layers that are all associated with specific security protocols [7].

The principal layer, Infrastructure as a Service, distinguish what sort of security threats roll in from outside the system.

The subsequent layer, Platform as a Service, is answerable for constant burden offsetting and planning with embedded systems.

The third layer, Software as a Service, is answerable for controlling the response between the safe vehicular cloud server and whoever is attempting to get to it, for example, vehicle proprietors and service providers.

Infrastructure as a Service (IaaS)	**Identify Units - Security Audit - Access Control - Storage Security**	**Vehicular Cloud Server Security Protocols**
	Cloud Storage Computing Data Analysis Virtualization/Standardization	
Platform as a Service (PaaS)	**Load Balancing Real-Time Scheduling Evaluation**	
	Real-time Data Processing Embedded System Storage Data Visualization	
Software as a Service (SaaS)	**Service API Interface Service Agent**	
	Interactive Browsing Analytical Computing Data Transaction	

12.3.2   Comparing the Current and a Connected Transportation System

In the current transportation system when an accident occurs on the road, communication about the accident to the family members, police station and hospital solely depends on the alertness of bystanders on the street or other drivers in the vicinity. If there is no one present in the surrounding area or no one is willing or capable of calling for help, then the parties involved in the accident are left without assistance example shows in Fig. 12.5.

In a connected transportation system after the accident occurs as shows in Fig. 12.6, there are internet connected devices such as the dashboard camera or the car's OBD and other nearby connected vehicles which can act as a source of news transmission to family members, police and medical assistance so that help can be alerted as quickly as possible [8].

In addition, in a connected transportation system, there would be no traffic jam because traffic would have been re-routed automatically as represents in Fig. 12.7.

12.4   Why Do We Need Online Driving Lessons?

It is a daunting task to teach someone to drive while sitting in a passenger's seat. As we see in the photo, the anxious reaction of the father responding to his son's reactions to upcoming traffic. Learning to drive with a person sitting in the passenger seat makes

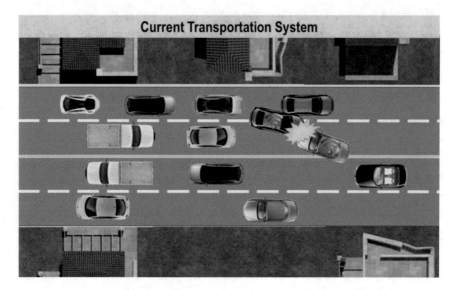

Fig. 12.5 Current transportation system

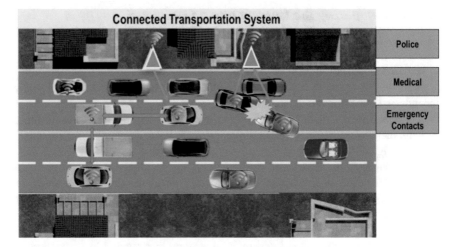

Fig. 12.6 Connected transportation system

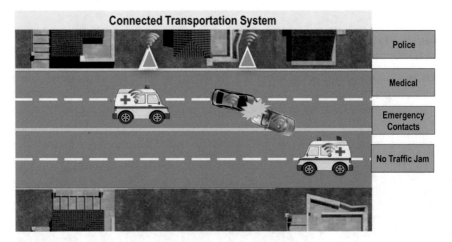

Fig. 12.7 Example of connected transportation

it a very difficult task because the person instructing might be providing guidance that your brain may not comprehend immediately and may cause a beginner to make mistakes and even panic. This is the reason why we need online driving lessons, which will be much less distracting than the traditional scenario. The OBD II, again, comes to the rescue. The data that is collected from the OBD II includes data related to the vehicle's health, data related to the driver's driving behavior and other important information. This information can also be used for training new drivers. The existing internal components coupled with new devices integrated in a new architecture can make this happen.

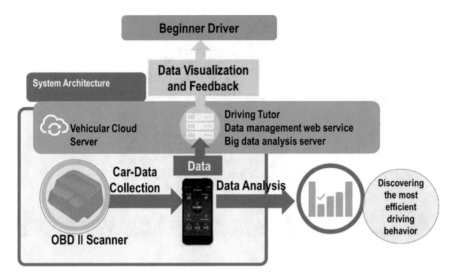

Fig. 12.8 Real-time driving system architecture

12.4.1   Real-Time Driving System Architecture

By utilizing a real-time driving system architecture shows in Fig. 12.8, there is a clear advantage: the beginner driver can get immediate feedback on the quality of his or her driving. The OBD II scanner provides information about acceleration, braking, and other information from the vehicle sensors to the smart phone via blue tooth. The smart phone conducts an analysis in real time utilizing the driving tutor application that is located in the vehicular cloud server. The data or results are visualized and presented as feedback to the beginner driver.

12.4.2   Discovering the Most Efficient Driving Behavior

New drivers usually drive their vehicles in an inefficient way; they step on the accelerator hard or "pump the gas" and use the breaks too often. All this information is part of the conventional data collected by the OBD II device. How is this data utilized to help discover the most efficient driving behavior?

The OBD II device collects data regarding engine load value, engine coolant temperature, short term fuel, fuel pressure, throttle position, etc. All this data can be inserted in various mathematical models. The two solutions depicted in the diagram provide the most efficient driving behavior. In below diagram, we see how the engine sound equalizer can be a good benchmark for throttle position and functioning. Various solutions can be developed for other kinds of vehicular performance captured by the OBD scanner, we can see in below Fig. 12.9.

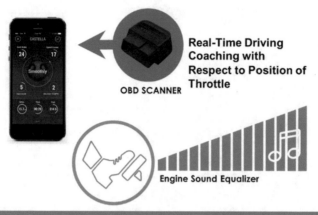

Engine Sound Equalizer for Better Understanding of Throttle Position and Functioning

Fig. 12.9 Efficient driving behavior example

12.4.3 Example: Efficient Driving Behavior Example or Looking for the "Sweet Spot"

We conduct a real-time throttle "sweet spot analysis" by using a road load power algorithm. We do this to discover how to reduce fuel consumption while on a trip, how much the throttle should be pressed to have optimum consumption in traffic and how to reduce the physical limitations of vehicle engine performance. The most efficient choice is known as the "sweet spot." In this diagram, the road load power algorithm—or sweet spot analysis is based on these parameters: linear acceleration, mass air flow, drag coefficient, pitch angle and rolling resistance as represents in Fig. 12.10.

12.4.4 Real-Time Coaching

All of this data from the sweet spot analysis can be used to coach drivers. In the new driver scenario, instead of the scared father trying to tell the driver how to drive, the real-time driving system architecture can communicate to the new driver real-time instructions from an app. The connected vehicle is equipped with feedback system capabilities that it can prove to be a much better coach. Algorithms that detect and find the sweet spots for something as simple as optimal fuel usage to more detrimental driving behaviors regarding safety will provide immediate feedback for every action taken by the driver.

In the below figure shown, notice the top bar of the screen showing the sweet spot analysis of "GOOD" or "BAD" to communicate if the driver is doing well at reducing fuel consumption and reducing engine load. Because this feedback is given

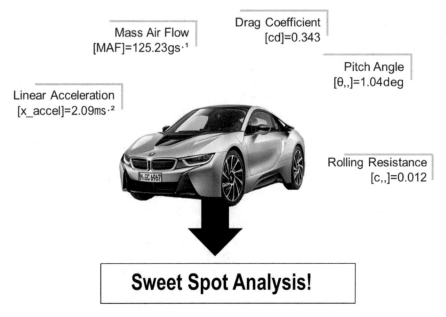

Mass Air Flow
[MAF]=125.23gs⁻¹

Drag Coefficient
[cd]=0.343

Pitch Angle
[θ,,]=1.04deg

Linear Acceleration
[x_accel]=2.09ms⁻²

Rolling Resistance
[c,,]=0.012

Sweet Spot Analysis!

Fig. 12.10 An efficient driving behavior example

in real time, drivers can adjust their driving behavior immediately which is likely to lead to permanent changes in driving behavior over time.

Fig. 12.11 Sweet spot navigation apps

A big part of driving feedback has to do with trip history. In Fig. 12.11, we examined real-time coaching and sweet spot navigation applications. Those provide formative real-time evaluation of the driver's overall behavior. By using trip history apps, the driver receives a summative evaluation of their driving behavior. These apps have features which can record the whole trip history and correlate it to engine performance. Trip History can also be useful for the car owner to keep track of the places they have visited, and system diagnosis can provide real-time vehicle health. Three examples of utilizing this feedback are trip history analysis for better economic driving, remote car-state diagnosis and car-repair shop recommendations and car supply management.

Driving applications provide an excellent opportunity for driving counseling and providing feedback analysis but are by no means a substitute for human supervision and interaction. They can be used by parents, instructors or other more experienced drivers in collaboration with the new driver to review the data and discuss the driving behaviors that are good and those that need adjustment. Many parents use these apps for monitoring their children's driving habits as shown in Fig. 12.12. The two underlining factors that keep coming up were safe and smart driving. By safe driving we mean the elimination of hard braking, hard acceleration, excessive speed, or sudden lane changes, etc. By smart driving we mean fuel cost reduction, right-time supply replacement, or faster navigation. Although these may sound simple concepts, but they are far from being achieved in the conventional vehicular universe as it stands today.

Fig. 12.12 Driving feedback and trip history

12.5   The "Smart" Parking Lot System

In the ordinary parking IoT system, there are a few issues that are amazingly irritating for drivers. In the first place, browsing without an organized parking system. At the point when drivers enter the parking IoT they do not understand in regard to by and large accessibility of parking spots and their area in this manner giving drivers trouble to discover a space. It is evaluated that the normal time expected to discover a spot is 20 min. Second, occasionally there are insufficient parking spots to oblige each vehicle entering the parking IoT so those drivers who show up after the expected time may locate no accessible space at all and simply waste their time. Time which they could have used to find an available space in another parking garage. Such scenarios often cause frustration and traffic inside a parking lot which affects incoming and outgoing vehicles. Third, it is estimated that 30% of traffic congestion in the world is due to the conventional parking systems. The smart parking lot has shown in Fig. 12.13.

In Fig. 12.14 has presented an example of a conventional parking lot map. All the vehicles depicted here are looking for a free parking spots as shown here in green blocks. Naturally, they must stop and check to see if there is an opening every so often. These stopping points are indicated by the red dots. In order to avoid all these stops, a smart parking lot would have a high precision indoor positioning system (Fig. 12.15).

The high precision indoor positioning system (Fig. 12.16) uses a device called sensor fusion which combines all the sensors: an accelerometer, a gyroscope, a magnetometer, a LIDAR sensor, a radar sensor and an inertial navigation system. In combination with a map of the parking lot, a high precision indoor positioning system can locate the position of vehicles inside the parking lot in real-time. A visual beacon attached near each parking space displays a guided path from the entry to the free parking space. In Fig. 12.17 shows the example of a smart parking lot map,

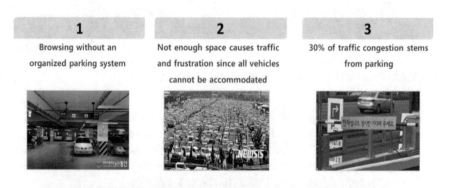

Fig. 12.13   Smart parking lot system

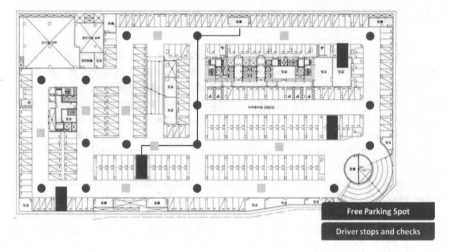

Fig. 12.14 Example of conventional parking lot map

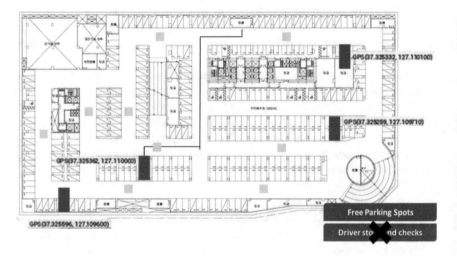

Fig. 12.15 Example of smart parking lot map

there is no need to stop and check for free spaces. Each space has a unique digital ID shown as GPS coordinates eliminating the need for the driver to stop and check for a spot. In a smart parking lot, a digitized the map and made it available to anyone using a specific app. This app will connect the location map to the vehicle's indoor navigation system and combine it with the last mile solution.

Here we see a mobile apps for a conventional parking lots verses a smart parking lot. The map at the far left shows us again that in the conventional parking lot the vehicle must spend 10–20 min to discover a free parking spot. Using the mobile app in a smart parking lot as shown on the right the free spots are indicated in the first

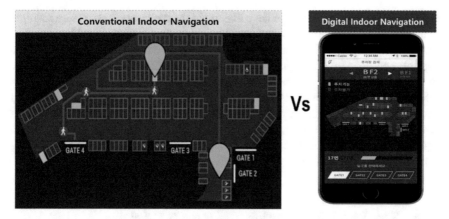

Fig. 12.16 Mobile apps for conventional versus smart parking lots

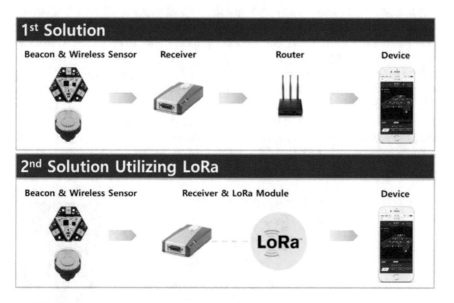

Fig. 12.17 Network architecture for smart parking lots

image and the nearest space and the optimal route to get to it is indicated in this app screenshot.

We can find the mobile apps for a conventional parking lots verses a smart parking lot in below fig. The map at the far left shows us again that in the conventional parking lot the vehicle must spend 10–20 min to discover a free parking spot. Using the mobile app in a smart parking lot as shown on the right the free spots are indicated in the first image and the nearest space and the optimal route to get to it is indicated in this app screenshot. In the next slide we will explain the necessary architecture required to make this solution possible.

Let us take a look at the network architecture for smart parking lots. These are possible solutions for developing a testbed setup regarding the prerequisite network. In the first solution, we see the beacon and other wireless sensors transmitting the information to the receiver via router then reaches a device which in this case is a smart phone. In an alternative architecture, we could replace the router with a Long-Range low power Wide Area Network module which in turn would communicate with the smart phone. The question that arises is what happens when the parking lot is underground, and the navigation system seizes to function?

12.6 The Last Mile Solution and How It Works

A smart parking IoT guide can end up being amazingly valuable for self-driving vehicles since they need a route framework to follow so as to leave. Such a guide can show the ongoing movement of the vehicle from the source to the goal (i.e., from the section of the leaving to the parking spot). Additionally, a keen parking IoT can be utilized to acquire real time, ideal path which can be utilized by either a self-driving vehicle or a driver driving the vehicle to offer security, precision and diminish traffic. This solution is known as "Last Mile Solution" as shown in Fig. 12.18. The main challenge addressed by the last mile solution is that the navigation service terminates when the vehicle goes underground. The smart parking lot's mobile app provides the last mile solution upon arrival and this service terminates after parking is completed.

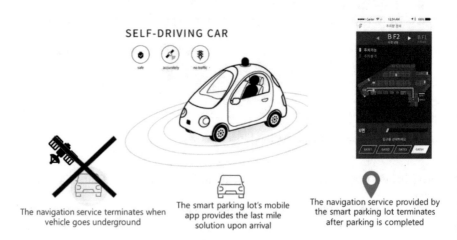

Fig. 12.18 The last mile solution and how it works

Fig. 12.19 Apps for underground smart parking lot solutions

12.6.1 Apps for Underground Smart Parking Lot Solutions

With the help of smart phone apps in smart parking lots, we can perform certain tasks. First, we can remotely access and control the underground parking lot beacons. Second, without going into the parking lot we can access the maps that are digitally available to anyone on the public platform and check where a vacant space is from the mobile phone. Third, we can see accurate location information regarding all the empty spaces in the lot. All of these technologies combined can help to minimize the time and energy spent to find the optimal parking space as well as costs. In the next and last slide, we will provide an overview of the whole system in Fig. 12.19.

12.6.2 An End-to-End Smart Parking Lot Solution

The entire application can be described as a one step, end-to-end solution for a smart indoor parking. The main challenge was to overcome the conventional problems of GPS limitations underground. Once that was solved by last mile navigation the road is open for implementation. In addition, because each vehicle has a unique ID that can be obtained by the OBD II, this approach can also solve for complex payment issues by replacing the existing conventional system with a mobile-based payment solution as shown in Fig. 12.20.

12.7 Summary

With smart parking application, this chapter has provided details to the need for an open platform, requirements of connected vehicles system structure, the required

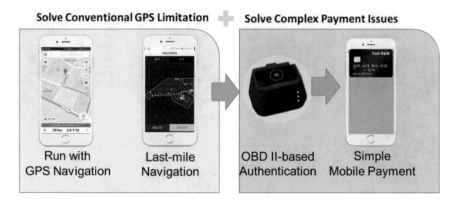

Fig. 12.20 Apps for underground smart parking lot solutions

connected vehicle system architecture, the vehicular open platform cloud server paradigm, advantages of connected vehicles, the OBD II as an automotive vehicle data collection source, automotive system architecture, secure vehicular cloud server platform, comparing the Current and a Connected Transportation System. We have also got the details of real-time driving system architecture, discovering the most efficient driving behavior, and also get the use case as an efficient driving behavior and Looking for the "Sweet Spot", and real-time coaching, the "Smart" Parking Lot System, last mile solution and process "How it Works", apps for underground smart parking lot solutions, and finally we have described an End-to-End Smart Parking Lot Solution.

Acknowledgements This research was funded by Woosong University Academic Research in 2021.

References

1. M. Gerla, L. Kleinrock, Vehicular networks and the future of Mobile Internet. Comput. Netw. **55**, 457–469 (2011). https://doi.org/10.1016/j.comnet.2010.10.015
2. M. Dikaiakos, S. Iqbal, T. Nadeem, L. Iftode, VITP: an information transfer protocol vehicular computing, in *ACM Workshop on Vehicular Ad Hoc Networks (VANET'05)*, Cologne, Germany (2005)
3. H. Hasrouny, A.E. Samhat, C. Bassil, A. Laouiti, VANet security challenges and solutions: a survey. Veh. Commun. **7**, 7–20 (2017)
4. Yan, S. Olariu, A probabilistic analysis of link duration in vehicular ad hoc networks. IEEE Trans. Intell. Transp. Syst. **12**(4), 1227–1236 (2011)
5. S. Olariu, M. Eltoweissy, M. Younis, Toward autonomous vehicular clouds. ICST Trans. Mob. Commun. Comput. **11**(7–9), 1–11 (2011)
6. C. Wang, Q. Wang, K. Ren, W. Lou, Privacy-preserving public auditing for data storage security in cloud computing, in *Proceedings of IEEE INFOCOM* (2010), pp. 1–9

7. G.-U. Rehman, A. Ghani, S. Muhammad, M. Singh, D. Singh, Selfishness in vehicular delay-tolerant networks: a review. Sensors **20**, 3000 (2020)
8. N. Magaia, Sheng, Z. ReF, IoV: a novel reputation framework for information-centric vehicular applications. IEEE Trans. Veh. Technol. **68**, 1810–1823 (2018)

Chapter 13
Adaptive Cyber Security Requirements for Intelligent Transportation System: A Case Study

Irish Singh and Madhusudan Singh

Abstract Intelligent Transportation System (ITS) is the emerging transportation which comprises of the advanced technologies such as wireless communications, video vehicle detection, distributed system architectures, human machine interface, sensing and actuating, to improve safety of the passengers, traffic congestion, fuel consumption and optimize other services of the transportation system such as real time traffic situation, electronic toll collection, automatic road enforcement and hot lanes. In order to build an ITS system having the capability to adapt itself, this paper presents the self -adaptive requirements for the ITS.

Keywords Requirements engineering · Self-adaptive · Intelligent transportation system

13.1 Introduction

Internet is developed following Person to Machine (P2M) paradigm which came from the first version of the distributed computing environment. User's request the server to get the service or data, and the server provide services or data based on the requested services. However, the IoT is now evolved to the Machine to Person paradigm (M2P), and then to the Machine-to-Machine paradigm (M2M) which are major model for future IoT beyond the P2M paradigm. In the perspective of M2P Model, machines will help people to support their work providing their services everyday every moment. Further, in the viewpoint of M2M, machines can communicate with other machine to provide the service to people without any interaction or intervene by human. In these kinds of model, machines are able to provide more services than machines in simple P2M model [1].

I. Singh
Department of Software and Computer Engineering, Ajou University, Suwon, Republic of Korea

M. Singh (✉)
School of Technology Studies, Endicott College of International Studies, Woosong University, Daejeon, Republic of Korea
e-mail: msingh@wsu.ac.kr

With the emergence of smart city research projects and other IoT related work, the Intelligent Traffic management system is rapidly evolving. With the cloud management of the IoT devices there is a shift of users to community centric approach from individualistic approach. The futuristic solutions to transportation system are appealing and beneficial for the automobile industry being the aim to improve the traffic condition by facilitating the vehicles with self-adaptive features, safety of people and cutting cost of ownership of an automobile. ITS is the urgent need of today [2]. According to the WHO records, the total number of road accidents is unexpectedly high, leading to 1.24 million per year [3]. Only 28 countries which represents 449 million people (7% of the world's population), have adequate laws that address all five risk factors (speed, drink, driving, helmets, seatbelts and child restraints). The WHO statistics records that there are 1.24 million road accidents and 31% of it is road accidents by only car occupants [4]. Intelligent transportation is the need of today's era. Upcoming advanced technologies like the IoT, wireless sensor network, cloud computing etc. can be used for monitoring the real time traffic data by the sensor devices and the OBD devices for prevention/detection of the accident. In order to build an efficient Intelligent Transportation system, we ought to have a clear set of requirements.

The Requirement Engineering Process
This section tells us about the Requirements Engineering and the advantages of the requirements engineering.

13.1.1 Requirements Engineering (RE)

Requirement engineering is a field of software engineering. Software engineering aims at designing, developing, delivering, and maintaining a piece of software. Requirement engineering deals with the very early phase of software engineering [5]. RE is all about communicating with the customer and understanding his/her problem. For a given work, we have to make sure that every requirement of the customer is covered, discussed and agreed. In requirement engineering, we can use many techniques to help in organizing, gathering, understanding the requirements of the system [6]. The aim of RE is to reach a full understanding of the problem domain, and to have a clear set of requirements that we can keep for the whole lifecycle of the work. We are supposed to communicate intensively with the customer, for that an agreed set of requirements means clear contract, clear identification of resources to allocate to the work and satisfaction of the customer. RE methodology, consist of 4 phases as shows in Fig. 13.1.

Context and Groundwork: In this phase, domain of the software system is defined. It takes a long time to analyze domain. Scope of the system is important and helpful for next three phases. In addition, proper RE methods and models are chosen. It includes elicitation, understanding and structuring, modeling and analysis as activities.

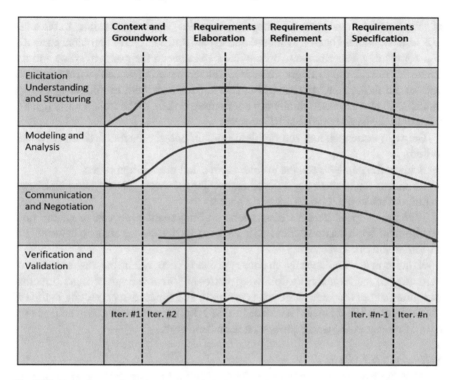

Fig. 13.1 Four phases of requirements engineering

Requirements Elaboration: In this phase, primary requirements are elicited. Also, Requirements in previous phase are redefined and come out due to stakeholders. It includes elicitation, understanding and structuring, modeling and analysis, communications and negotiations, verification, and validation as activities.

Requirements Refinement: In this phase, requirements from the last phase are refined due to the system model. The system model is defined in a detail. In this phase, communications and negotiations with stakeholders are more focused.

Requirements Specifications: In this phase, requirements specifications, which are the description of the behavior of the system to be, are defined. Verification and validation are more focused on rather than other phases.

13.1.2   Advantages of Requirement Engineering

The answer to the question that why so much time is invested in the RE phase and not directly going to the Analysis Phase or Implementation Phase is given through the advantages of RE. The development of any system needs a thorough understanding of "What problem the system is supposed to solve?" So, before rushing to the analysis or development phase we should understand and analyze the problem correctly and completely. Requirements bridge that gap between the problem space and the solution space. Through a rigorous requirement engineering process, we can not only understand the stakeholders needs but also can analyze them by following various models and processes and finally can communicate them to the stakeholders [6]. So, to sum up the basic benefits of RE process:

Helps in understanding the problem domain and the intended behavior of the system.

Bridges the gap between the problem space and the solution space.

Provides various methodologies to analyze the behavior of the system which helps to find out the conflicts at a very early phase.

With the help of iterative discussions and negotiation we can reach the final requirement specification which will be used as formal agreement between the customer and the developer.

Requirements engineering, if done properly can minimize the expense of correcting a requirement in the following phases of Software Engineering. Correcting a requirement in the later phases can cost up to 200 times than correcting in the RE phase (Boehm 1981; Davis 1993; Buede 1999). RE is the phase that will help us save a lot of money in the whole process; it is an "enabler".

Self-adaptive System

A self-adaptive system works in a different way; it first analyzes the environment, senses the environment and then takes decision that fulfills its requirement. The environment of the system is very complex so optimal decisions cannot be taken but a good enough decision is taken which fulfills the requirement of the system. A self-adaptive system is aware of its architecture so it can reconfigure [7].

13.2 MAPE-K Model

MAPE-K stands for Monitor, Analyze, Plan, Execute and Knowledge. Self-Adaptive Systems use this activity loop for adaptation represents in Fig. 13.2.

MAPE-K has four processes and one knowledge base.

Monitor: Monitor phase monitors the environment and itself and uses scenario to collect basic data.

Analyze: This is the second phase, and, in this phase, the basic data is analyzed and understood and given meaning.

Plan: The third phase of the activity loop is Plan. In this phase a plan is created by selecting sequence of actions to be conducted in the understood environment.

Execute: The fourth phase is Execute. In this phase, the plan is executed via the actuators (effectors) of the system, and the environment is modified. The monitoring

Fig. 13.2 MAPE-K
activities

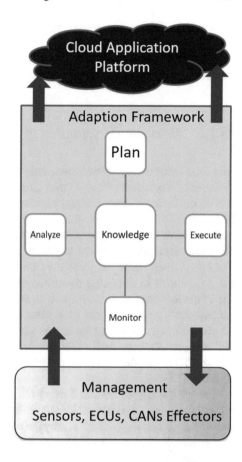

activity will detect potential changes, and the system will keep performing the 4 activities. Knowledge is the knowledge base of the system. The planning activity is using the knowledge of the environment.

13.2.1 Self-adaptive Intelligent Transport System

The environment of the system i.e., the Intelligent transportation system is very unpredictable and changeable as in the environment there will be humans driving the car, pedestrian, moving on the pathway so human's nature could be unpredictable, and some mistakes can be made by them. Therefore, the system should sense and analyze the system all the time and accordingly abstract the important information and plan accordingly [8]. For example, if someone breaks any traffic rule and creates nuisance on the road then the system suspects the person breaking rule or creating any nuisance as obstacle and alerts other drivers in the range of the unexpected event.

The systems main requirement is "Safety", therefore, the other drivers can change the path of their driving or may break some traffic rule while getting the alert message, but the system's critical requirement of "safety" has to be fulfilled, even though some requirement like "not compliance with traffic rule" can be compromised [9].

13.2.2 Domain: Presentation of Domain

The Intelligent transportation environment uses many technologies such as WSN (Wireless Sensor Networks), Cloud computing, Networking and Communication and so on further. Smart vehicles consist of smart monitoring and control devices like OBD (On Board Diagnostic), Wi-Fi and Bluetooth enabled driving assistance system and other plugins connected with your smart mobile phones, these devices share the information (accident, traffic etc.) with another user through ad hoc network. The smart device will just share the data i.e., video, audio, location info. via GPS, etc. to the user in the range of ad hoc network or to a cloud server using Wi-Fi. These devices consist of modules to accumulate and pass data and information like camera modules for rear and front view, LCD module, GPS Module, and different sensors like pressure sensor for accident detection, temperature sensor for protecting the data, proximity sensor for vehicular lane change detection, and speed sensor [10]. Figure 13.3 shows the overview of a smart device enabled Intelligent Transportation System (ITS).

The information is collected using the collectors, the front and rear camera for recording the pre- and post-accident data. The information gathered is sent to the end users after processing who are nearby the ad-hoc network third party not. In the event of accident, the video, clips recorded, location is sent to a cloud server to inform the authorities like hospital, police station fire brigade, etc. [6, 7].

Figure 13.4 shall explain the features of the development platform for Intelligent Traffic Management System. It shall collect information about the traffic from the crowd (Crowd Sourcing), the running automobiles on the road, as well as the weather conditions from weather information Centre [11]. The smart device will make use of the info. and communicate the information to the nearby vehicles and the authorities.

User Characteristics: A user is one who has the legitimate qualification of a driving, i.e., user can drive, know traffic signs and able to figure out traffic signs, other vehicles signs and traffic laws. Users must follow the system's guidelines. User is not involved in any malicious activities. Users of the system are the authorities such as the hospital, Police station, user's family.

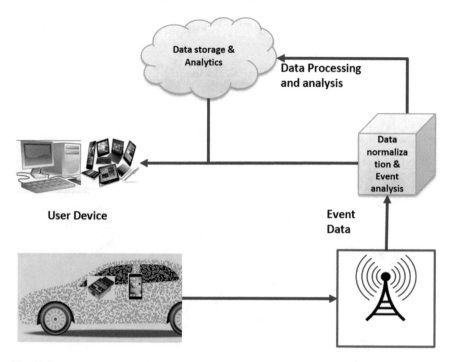

Fig. 13.3 Intelligent transportation systems

13.3   Related Work

13.3.1   Cloud Computing Based Urban Traffic Control System

This paper proposes a traffic control system whose goal is enhancing the throughput of road and optimization of the traffic control for the safety of the participants by using Intersection Control Services which are a part of the city/region wide cloud system that coordinates flow of traffic between intersections. They use the geographical multicast address for targeting all vehicles in that particular region. The ICS gathers the traffic data from various sensors around the intersections and from the vehicles for short term prediction for vehicle control. The vehicles are treated as services by the system perspective and these vehicles are discovered and invoked by cloud methodology [2].

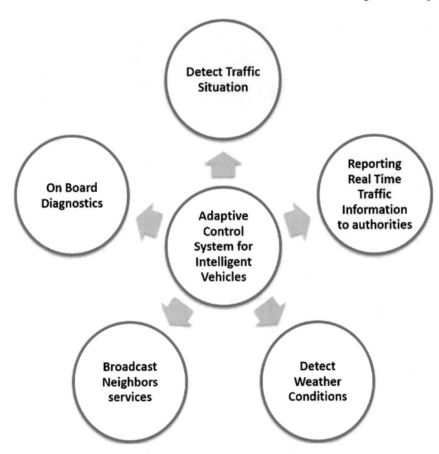

Fig. 13.4 Features of adaptive control system for intelligent vehicles

13.3.2 CloudThink: A Scalable Secure Platform for Mirroring Transportation System in the Cloud

In this research work which have been done by Erick Wilhelm et al. [3] they developed a platform for vehicle communication consisting of low-cost open-source hardware CARduino to move the vehicle data to a secure server, an API for providing third party services Infrastructure for Interacting with users and a dashboard for access control and service distribution. The CloudThink infrastructure promotes the commoditization of vehicle telematics data by facilitating easier, flexible, and more secure access. It enables drivers to confidently share their vehicle information across multiple applications to improve the transportation experience for all stakeholders, as Ill as to potentially monetize their data [3].

13.3.3 Crowd ITS: Crowdsourcing in Intelligent Transportation Systems

Kashif Ali et al. proposed, CrowdITS, Crowdsourcing in ITS. Human inputs, along with available sensory data, are collected and communicated to a processing server using mobile phones. The basic idea is to use the Crowd with smart mobile phones to enable certain ITS applications without the need of any special sensors or communication devices, both in-vehicle and on-road [4].

13.3.4 ITS-Cloud: Cloud Computing for Intelligent Transportation System

In this paper a new cloud computing model called ITS-Cloud is applied to the Intelligent Transportation Systems (ITS) to improve road safety, transport productivity, travel reliability, informed travel choices, environment protection, and traffic resilience. The proposed system consists of two sub-models: the statistic and the dynamic cloud sub-models. In the statistic model, vehicles benefit of the conventional cloud advantages; however, the dynamic one which is a temporary cloud is formed by the vehicles themselves which represent the cloud datacenters a simulation study is performed to deal with the load balancing as a NP-Complete problem. The reached results are obtained using Bees Life Algorithm (BLA) applied to ITS-Cloud and compared with those reached by (BLA) applied only to the conventional Cloud [5].

13.4 The Requirements Engineering Process Applied to Its

Phase 1: Context and groundwork, requirement elicitation
We performed a review of existing ITS and made preliminary scenarios. As the system needs to be self- adaptive, we studied how requirements engineering is different in self -adaptive systems. A goal-based approach is usually used. We identified actors and goals from the scenarios.

Some of the scenarios are listed below:

Bold Text: Actor.

Underlined Text: Goal.

Scenario 1: Car moving on the highway.

No problem.

The driver is moving on the highway towards his destination. The driver gets the traffic information, weather information from Wi-Fi devices on the intersections. The driver also gets the information of additional services like golf outlet, amusement park etc. with ongoing benefits (The discounts going on various services) via interface.

Accident happens within the range of Ad-hoc network of the user
The system alerts the driver about the accident. System notifies the user verbally and visually, the details of the accident, like location of accident, time of accident and the current status of the traffic at the location of the accident from the intersection as well as from the cars in the range of ad-hoc network of the accident. The system also suggests alternate route to the destination in case of congestion on the current route.

In front car of the user met with an accident
The system automatically slows down the speed of the car and alerts other vehicles in close proximity of the user, about the accident. The system collects the accident information, like the pre- and post-accident video, audio, pictures, location, time and sends the accident data captured to the nearby intersection device and to other cars in close proximity of the user and in the range of the ad-hoc network.

The car itself met with an accident
The system automatically captures all the information from the OBD i.e. status of all the sensors, the pre and post-accident pictures, video, audio clips, owner's car number, name, mobile no., address and other important details and sends it to the intersection device and other vehicles in the range of the ad-hoc network.

Other scenarios are listed in Appendix.

Phase 2: Requirements elaboration and analysis
The second phase of requirement engineering is Elaboration. In the previous phase, we have naturally elicited some requirements from our documentation. As our work was not structured, we know that we have certainly missed some core requirements. In this new phase, we refine our primary requirements by connecting scenarios to goals in a single goal tree model which we refine until we find our requirements, as leaf goals. In order to find a comprehensive set of requirements, we have decided to use the goal modeling KAOS approach.

13.5 Usefulness of Goal Modeling in Requirement Engineering

A goal is an objective that the system under consideration should achieve. So goal formulations are very important to refer to the intended properties of the system.

Through goal modeling we can intrinsically characterize each goal by name or specification. We can also decompose each higher-level goal to get lower level sub goals and eventually we can operationalize each of them. Operationalized goals can lead us to the specific requirements which were not so comprehensible before.

If right methodology is chosen for goal modeling, it can help us linking not only the requirements but also the goals with the agents.

We can also identify various types of goals, for example Maintain (goal that needs to be met all the time), Achieve (Goal that needs to be reached eventually), Optimize (Goal that ensures some soft target property). In our work, we have not explored this possibility, but this is one potential improvement. This characteristic is particularly useful when we want to reason about the satisfaction of high-level goals, which can be formally decomposed into combinations of satisfactions of the sub goals depending on their type. In our case, we focus on requirements only (leaf goals) so the use of goal types is not significantly useful.

Moreover, goal modeling is one of the best ways to analyze and discuss about the potential conflicts as it gives us a big picture of the system and its intended behavior.

13.5.1 Existing Approaches for Goal Modeling

Several approaches have already been designed to address Goal Modeling for requirement engineering. In this report, we will point out three of them, and we will explain why we have chosen the KAOS approach. GRL [11] (Goal Requirement Language) is a language which aims at exploring the connections between various actors. It uses three kinds of concepts: intentional elements, links and actors. Basically, these concepts permit to connect the stakeholders together by showing the rationale of their relationships: for instance, one actor depends on another for the access to a certain resource. In addition, classic notations such as hard goal, soft goal or contribution links are available. i* [12] is the "child" of GRL. It extends GRL and its use becomes closer to agent-oriented design. Through, the use of two diagrams, the designer can elicit and obtain a well-defined and structured understanding of the intentional concepts in a system. The Strategic Dependencies (SD) model shows how actors are connected to each other through dependency links, while the Strategic Rationale (SR) model shows how an actor can fulfill his goals. KAOS [13] is a requirement engineering method that permits to elicit the requirements of a system thanks to a goal modeling approach. The designer will elicit the goals of the system, and through a top down or a bottom-up approach, s/he will finally obtain a set of leaves goals that are said to operationalize the previous goals, thus becoming the requirements of the system to be. The requirements are taken over by agents (or software entities). The goal model permits to link requirements to soft goals, to visualize conflicts between goals and to elicit obstacles that have a negative influence on our requirements or goals [14].

13.5.2   Motivation to Choose KAOS

We have selected KAOS for various reasons. One reason is that KAOS is relatively mature since it has been designed 20 years ago. KAOS is still being used and is a reference in terms of Requirement Engineering. This is a method that has proven itself to be consistent and useful. KAOS is also used in various industry projects, and a tool (Objectiver [15]) is still actively developed and sold. Moreover, there are tutorials available for KAOS [10] to help beginners. In addition, KAOS is quite simple to use and more intuitive than i* for instance. It also provides every basic feature that we need to use in the RE process and is quite easy to extend if we need to for our specific domain.

13.5.3   Constructing the Goal Model and Identifying Hard/Soft Goals

Our main objectives of constructing a goal model are: To reach right requirements which are easy to comprehend and validate and to introduce right level of flexibility in the system and allowing self-adaptiveness.

- **Determining hard and soft goals**

    We identified the hard and soft goals from the rough requirements which were elicited in the last phase. We mainly used the scenarios, where actors and goals were defined. We categorized those goals as hard goals when their satisfaction could be established through verification techniques in a tangible manner: they are measurable. In contrast, the goals whose satisfaction cannot be established in a clear-cut way are categorized as soft goals.

    Actually, we have elicited many different soft goals (Non-Functional Requirement, quality requirement) in the next phase (requirement specification), helping ourselves with the design criteria tables. An explanation will be given in the next section of the report. We did not include the soft goals in the model itself, but they are mapped to hard goals to preserve traceability and allow conflicts resolution.

- **Decomposing the high-level goals to sub goals to reach requirements**

    Firstly, we made a topmost goal and started decomposing it by asking "How" questions till we reach the goals which are operationalizable [15]. Instead of following a strict 'top down 'or 'bottom up' approach of goal modeling, we analyzed scenario as well as our elicited requirements to construct the goal model. By doing this we could avoid the curse of being too much specific or too much general. Our main objective of this goal modeling was to come up with the requirements and at the same time to cover almost all possible scenarios.

    We defined our topmost goal as 'Destination reached in safe, efficient and comfortable way using Intelligent Transportation System' which can be achieved by meeting

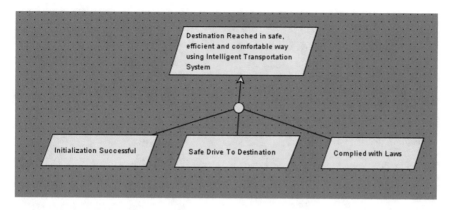

Fig. 13.5 Topmost goal to next sub-goals

3 next level sub goals 'Initialization successful', 'safe drive to destination' and "Complied with law". 'Initialization successful' goal is concerned with all kind of activities that need to be done before starting the vehicle. For example, On Board diagnosis, getting required information from user [16].

The next is the vehicle should drive safe on the road and follow the point-to-point route to reach the destination which is taken care by the goal 'safe drive to destination'. Finally, the system must comply with the law and traffic laws [17] as represents in Fig. 13.5.

We further decomposed each of these identified sub goals to reach a right level where the goal can be operationalized and linked to requirements. For example: 'Initialization successful' can be achieved by completing the On-Board diagnosis, informing the user about the status of the diagnosis and getting the destination specified by the user. We attach in annex our full goal model. As we said before, the final requirements for our system are the leave goals, which can be written directly in the software specification document.

For instance, in this Fig. 13.6, the goal On Board Diagnostic can be broken down in 4 sub goals, all of them being specific enough to be considered requirements (with the bold outline).

13.5.4 Identifying Uncertainty

Incidents in the environment and providing countermeasures as presents in Fig. 13.7.

We also analyzed the obstacles that the system might encounter at runtime and provided the countermeasures for them [18]. For example, when we specify the destination, the system is supposed to accept the destination but maybe there is no such address available in the GPS, or at least the system is unable to detect them. In such case, we decided that the user should be allowed to renter the destination.

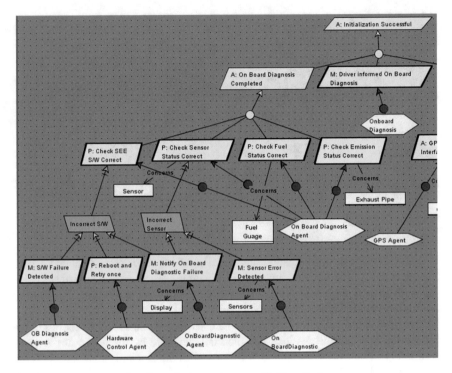

Fig. 13.6 Breaking a goal until we obtain requirements (Bold outline)

We have added another feature to our goal model which permits to introduce more adaptiveness into our system. For each goal, we define a temporal constraint. That is, we say that a goal has to be satisfied: Instantaneously: this is a perform goal. Until some state is reached: this is an achieve goal All the time: this is a maintain goal. The differences between these goals are very important for an adaptive system. Thus, for each goal or requirement we defined a goal type.

13.5.5 Identifying the Agents

After we elicited all our requirements and completed our goal tree, we decided to classify in term of their roles in the system. We tried to find similarities between goals and to assign agents to them. These agents are responsible for the conduction of the actions necessary for fulfilling the goals. We can then identify necessary subsystems and encapsulate behavior or functionalities in different modules. For each requirement, we found an agent to assign it to as we can see in Fig. 13.8. Then, we simply gathered all the requirements that the agents are responsible for and created responsibilities diagrams, which are part of the KAOS model.

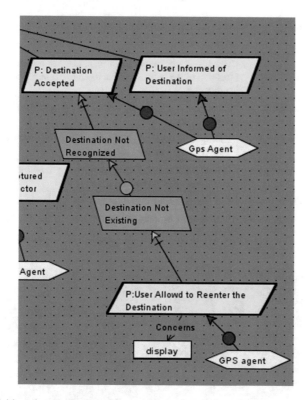

Fig. 13.7 Eliciting obstacles and providing countermeasure

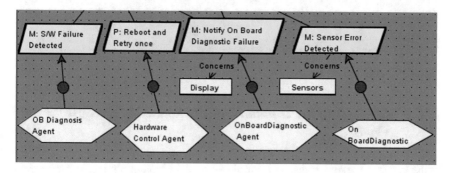

Fig. 13.8 Assigning agents to requirements

In this example, we identify two agents for 4 requirements. After assigning one agent for each requirement, we obtain a responsibility diagram. We identified a total of 6 agents. Table 13.1 has represents the Agent name and Description of their works. Figures 13.9 and 13.10 show the on board dignostic and hardware control agents.

The responsibility diagrams can be found in the appendix.

Table 13.1 Showing Agent's name and its corresponding responsibilities

Agent name	Description
GPS Agent	This agent is responsible for managing the information about the location of the vehicles in the route to the user and other authorities
System Collector Agent	This agent is responsible for collecting the sensor information, audio, video
Hardware Control Agent	This agent is responsible for controlling the hardware of the system like automatic brakes
Onboard Diagnostic Agent	This agent is responsible of maintaining the sensor, fuel and engine information of the vehicle
Law Management Agent	This agent is responsible for maintaining traffic laws all the time
Environment Context Agent	This agent is in charge of maintaining updated information about the external environment

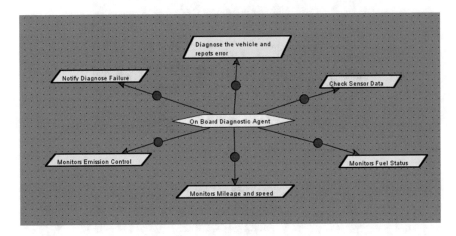

Fig. 13.9 On board diagnostic agent

13.5.6 Identifying the Domain Objects Related to the Soft Goals

In the last step we map objects of our object model (domain model) to the requirements. It is very helpful for a SAS, because the controller of a SAS has to monitor some variables [19]. Thus, we know which object is attached to what requirement.

We show an example below:

Phase 3: Requirements specification
After we elaborate and analyze the elicited requirements, we make the Requirement Specification which is the description of the behavior of the system to be. This document included the functional, as well as nonfunctional requirements and constraints

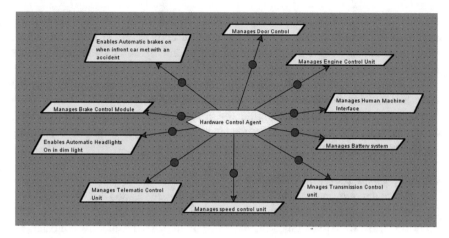

Fig. 13.10 Hardware control agent

of the system. Requirements specification permits a rigorous assessment of requirements before design can begin and reduces later redesign [20]. Finally, this document establishes the basis for agreement between customers and contractors or suppliers on what the software product is to do as well as what it is not expected to do. In Fig. 13.11 presents the user allowed re-entering the destination and its related object.

We focused on four main things while writing the Requirement Specification:

The leaf nodes of the goal models are considered to be as requirements as those leaf nodes are operationalizable and are assigned as a responsibility of a single agent.

We have identified those requirements which can be relaxed at runtime for the changeable environmental conditions. Then we have written those requirements using RELAX which is a language for writing the adaptive requirements. But, as

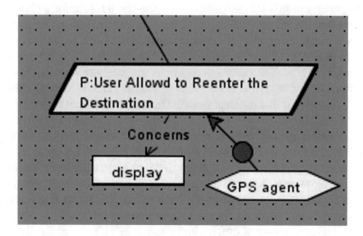

Fig. 13.11 User allowed re-entering the destination and its related object

our system is a safety critical system, we did not opt for relaxing most of the requirements. Only those requirements, which are not directly concerned with the safety, have been relaxed. We have also identified corresponding monitor able properties of the environment and how to monitor or sense them.

We have also provided a table which specifies which agent is accountable for what requirement. This would enhance the system's testability.

We have also specified the nonfunctional requirements and the corresponding design criteria. We explicitly document the external requirements or constraints imposed on the system.

Phase 4: Verification and validation

The specified requirements from the previous phase were verified that they are unambiguous, no redundant, complete, understandable, consistent and traceable [21]. Some problems were found, mostly relating ambiguity and traceability. For example, we found that in FR1, "test script" was ambiguous. It is not defined exactly what this test should do, so the change planned was making it clear what this script must at least contain. FR9 and F10, the statement about destination specification by voice recognition and touch pad, were not traceable to the scenario.

For validation, we made a scenario and checked that our requirements support it. Problems were written down and changes were planned.

The scenario: Mr. 'A' enters his car and starts the Car. The On Board Diagnostic diagnoses the vehicle; it checks the working of different sensors emission control, mileage, and speed and fuel status and if there is some problem with the sensor, the Interface informs the problem to Mr. 'A' with the message that the 'car needs maintenance'. If the OBD notifies diagnose failure the system reboots the OBD in case of failure and reports errors if any. If there is an error during startup of the software, the system informs Mr. 'A' and tries to reboot once. If it does not solve the problem, automation is disabled until maintenance is carried out. Mr. 'A' is moving on the highway towards his destination. He gets the traffic information, weather information and the information of additional services like golf outlet, amusement park etc. with ongoing benefits (The discounts going on various services) via interface from Wi-Fi devices on the intersections. Accident happens within the range of Ad-hoc network of Mr. 'A'. The system alerts Mr. 'A' verbally and visually, the details of the accident, like location of accident, time of accident and the current status of the traffic at the location of the accident from the intersection as well as from the cars in the range of ad-hoc network of the accident. The system also suggests alternate route to the destination in case of congestion on the current route and the system collects the accident information, like the pre- and post-accident video, audio, pictures, location, time and sends the accident data captured to the nearby intersection device and to other cars in close proximity of the user and in the range of the ad-hoc network. The data is processed at the device on the intersection. The server at the intersection analyzes the data and sends the data in real time to authorities like hospital, police station, fire brigade (in case of fire) for investigation and also to the family members of the injured person in the accident.

13.6 Requirement Specification

13.6.1 Functional Requirements (FR)

FR1: **As part of the initialization process, the system SHALL check the functionality of the software by running the test script. [The test script will be like simulation of random travel].**
  Related NFRs: Reliability **[completeness]+**, Testability **[accountability]+**
  Responsible**: On Board Diagnosis Agent**
  FR2: **In case of ITS software failure, the system SHALL detect the failure and notify the user.**
  Related NFRs: Reliability+, **Usability [communicativeness]+**
  Responsible: **On Board Diagnosis Agent**
  FR3: **As part of the initialization process the system SHALL check whether the sensors are functioning well or not. [The sensors SHALL provide right answers on request for testing].**
  Related NFRs: **Reliability [completeness]+**
  Responsible: **On Board Diagnosis Agent**
  FR4: **In case of sensor failure, the system SHALL detect the failure and notify user.**
  Related NFRs: Reliability+, **Usability [communicativeness]+**
  Responsible: **On Board Diagnosis Agent**
  FR5: **At any point of time, the system SHALL make sure that there is enough fuel available to reach the nearest gas station or to drive minimum of 30 km.**
  Related NFRs: Reliability [completeness]+
  Responsible: On Board Diagnosis Agent
  FR6: If the available fuel is not enough to reach nearest gas station or to drive minimum of 30 km, then the system SHALL notify the user.
  Related NFRs: Reliability [completeness] + , usability [communicativeness] +
  Responsible: On Board Diagnosis Agent.
  FR7: As part of initialization process, the system SHALL notify the user about the progress and the final result of the On-Board Diagnosis (software, sensor, fuel) in a user-friendly format.
  Related NFR: Usability [communicativeness]+
  Responsible: On Board Diagnosis Agent
  FR8: The system SHALL capture the destination from GPS interface.
  Responsible: GPS Agent
  FR9: The system SHALL determine whether the provided destination is reachable or not.
  Related NFR: Reliability+
  Responsible: GPS Agent
  FR10: In case of unavailability of destination, the system SHALL notify user and SHALL allow the user to reenter the destination.
  Related NFR: Usability [communicativeness]+

Responsible: GPS Agent

FR11: The system SHALL notify the user about the acceptance of the provided destination.

Related NFR: Usability [communicativeness]+

Responsible: GPS Agent

FR12: The system SHALL sense weather information like temperature, humidity, precipitation, wind speed, adherence of the road.

Responsible: Hardware Control Agent

FR13: If a collision is detected then system SHALL inform the accident information to the nearby vehicle by V2V connectivity and also to the intersection.

Related NFR: Reliability+, Usability

Responsible: System Collector Agent

FR14: The system SHALL notify the user about the services available in the route with their currents offers or discounts based on the GPS.

Related NFR: Usability+

Responsible: GPS Agent

Nonfunctional Requirements (NFR)

Requirements and Corresponding Design Criteria:

Reliability requirements:

NFR1: The system shall be operating without direct access to the network for a period of time (self-contentedness).

NFR2: The detection of the accident shall be accurate: the position of an object shall be detected with an error range not more than a measure that is function of the distance and the speed between the object and the vehicle (accuracy).

NFR3: The system shall not start if any of the necessary components is not available or not functioning (completeness).

NFR4: The system shall provide a formal semantic for any information that is used by the reasoning engines. For instance, ontologies might be used to describe the self and external environment, as well as the laws (consistency).

NFR5: The system shall be capable of performing normal operations under severe meteorological conditions such as ice rain, heavy snow, slippery roads, heavy mist (robustness).

Usability requirements:

NFR6: A function of the system shall always be accessible in less than 3 steps for the driver (accessibility).

NFR7: The system shall provide all the notifications created by the control software to the user in natural language (nontechnical ways) (communicativeness).

NFR8: The system shall provide the user with a minimum number of required interactions: the user only has to initialize the system in less than 5 min (I/O Rate).

13.6.2 Testability Requirements

NFR9: The system shall map the different subsystems/components to clear responsibilities, so that in case of testing/failure the error shall be easily located, in this case, the system could be aware of its own rationale, keeping a model of its goals and their relations to requirements (accountability).

External requirements:

NFR10: The system shall respect the law all the time, except when the safety of the passengers or other road users is at risk. That is, the system shall be aware of the traffic laws all the time, at any place (Legal Constraint).

Conflict: This requirement (NFR10) is conflicting with NFR1: if the network were always available, up to date information will be available at runtime without a need for storage. But because we relax the need for constant connectivity in NFR1, we need to agree on a compromise: the system shall save data about the laws changing with regard to region within 50 km.

Evolution, Maintenance and Traceability
In order to achieve an end-to-end traceability in the RE process, we update a table which keeps track of the Requirement ID, corresponding high-level goal and the concerned Scenario ID. It helped us to do the following activities:

Can easily identify the source of the requirement.

Can find out the main objective/purpose of the requirement.

If some requirement conflicts with other requirement we can modify one of them without hurting the concerned high-level goal/objective.

If some scenario is modified later, we can easily trace and find what requirements are now subject to change. This helps a lot in the maintenance and evolution of the requirements. Table 13.2 has represents the Requirement ID, each id priority and Goal with the Scenario ID (Table 13.2).

13.7 Conclusion

In goal modeling, we decompose business level goals (abstract, not operationalized goals) into sub goals until we find a goal that is concrete enough to be seen as a requirement: we say that it is operationalized; that is, its fulfillment can be measured, verified. Of course, we can follow top down, bottom up or hybrid approaches. Thus, when we are designing our goal model with KAOS, we want to stop breaking goals into sub goals when we see that a goal can be taken over by an agent (a subsystem/ module of our software). In the paper we could have used the viewpoint-based approach more to get a better view on the different perspectives on the system. We could have also considered about the security requirements and privacy requirements when sending

Table 13.2 Traceability table

Requirement ID	Priority	Goal	Scenario ID
FR1	High	Initialization successful (on board diagnosis)	1.4
FR2	High	Initialization successful (on board diagnosis)	1.3, 1.4
FR3	High	Initialization successful (on board diagnosis)	1.1, 1.2
FR4	Medium	Initialization successful (on board diagnosis)	1.2
FR5	High	Initialization successful (on board diagnosis)	1.1
FR6	High	Initialization successful (on board diagnosis)	1.1
FR7	High	Initialization successful (on board diagnosis)	1.1, 1.2, 1.3, 1.4
FR8	High	Initialization successful (destination specified)	2.1, 2.2
FR9	High	Initialization successful (destination specified)	2.1
FR10	Medium	Initialization successful (destination specified)	2.1
FR11	High	Initialization successful (destination specified)	2.1
FR12	High	Drive safe (weather info.)	2.1
FR13	High	Drive safe (avoid obstacle)	2.2, 2.3, 2.4
FR14	Medium	Drive safe (benefit services)	2.1

the accident information on the cloud server or V2V communication. This could be one of the future works of this article.

Acknowledgements This research was funded by Woosong University Academic Research in 2021.

References

1. World Health Organization, Global Status report on road safety 2015. Available: https://www.who.int/violence_injury_prevention/road_safety_status/2015/en/
2. X. Xu, W. Wang, Y. Liu, X. Zhao, Z. Xu, H. Zhou, A bibliographic analysis and collaboration patterns of IEEE transactions on intelligent transportation systems between 2000 and 2015. IEEE Trans. Intell. Transp. Syst. **17**(8) (2016)
3. P. Jaworski, et al., Cloud computing concept for intelligent transportation systems, in *14th International IEEE Conference on Intelligent Transportation Systems*, Washington, DC, USA, 5–7 Oct 2011
4. B. Okumura, M.R. James, Y. Kanzawa, M. Derry, K. Sakai, Challenges in perception and decision making for intelligent automotive vehicles: a case study. IEEE Trans. Veh. **1**(1) (2016)
5. F.Y. Wang, Driving into the future with ITS. IEEE Intell. Syst. **21**(3), 94–95 (2006)
6. D. Singh, M. Singh, I. Singh, H.-J. Lee, Secure and reliable cloud networks for smart transportation services, in *The 17th IEEE International Conference on Advanced Communication Technology (ICACT2015)*, Phonix Park, South Korea, 1–3 July 2015, pp. 358–362
7. T. Nishi, D.P.E. Wilhelm et al., Cloudthink: a scalable secure platform for mirroring transportation systems in the cloud. Spec. Issue Smart Sustain. Transp. **30**(3), 320–329 (2015)
8. K. Ali, et al., CrowdITS: crowdsourcing in intelligent transportation systems, in *2012 IEEE Wireless Communications and Networking Conference: Services, Applications, and Business*

9. S. Bitam, et al., ITS-Cloud: cloud computing for intelligent transportation system, in *Globecom 2012—Communications Software, Services and Multimedia Symposium*
10. J.J. Blum, A. Eskandarian, L. Hoffman, Challenges of intervehicle ad hoc networks. IEEE Trans. Intell. Transp. Syst. **5**(4), 347–351 (2004)
11. GRL was developed by the university of Toronto (including Eric Yu, the father of i*). Here is the official page of their research team. https://www.cs.toronto.edu/km/GRL/
12. E.S.K. Yu, Towards modelling and reasoning support for earlyphase requirements engineering, in *Proceedings of the Third IEEE International Symposium on Requirements Engineering, 1997*. IEEE, 1997, pp. 226235
13. D. Singh, M. Singh, Internet of vehicles for smart and safe driving, in *International Conference on Connected Vehicles and Expo (ICCVE)*, Shenzhen, 19–23 Oct 2015
14. This tutorial is the official tutorial made by respect IT, the team from University of Louvain who maintain and develop the KAOS methodology nowadays. https://www.objectiver.com/fileadmin/download/documents/KaosTutorial.pdf
15. Objectiver is a tool developed by the University of Louvain Belgium. It uses the KAOS methodology, which was developped by the cooperation between the University of Oregon and the University of Louvain in 1990. The tool is developped by respect IT, a spinout company of the University of Louvain
16. M. Singh, D. Singh, A. Jara, Secure cloud networks for connected & automated vehicles, in *2015 International Conference on Connected Vehicles and Expo (ICCVE)*, pp. 330–335 (2015). https://doi.org/10.1109/ICCVE.2015.94
17. R. da Rosa Righi, A.L. Santana, C.A. da Costa, R. Kunst, G. Goldschmidt, D. Kim (Singh), M. Singh, Reducing cost and time-to-market on supporting driver assistance systems to avoid rear-end collisions in vehicles traffic, August 1–3, 2019, in *IEEE International Conference on Computational Science and Engineering (CSE), and IEEE International Conference on Embedded and Ubiquitous Computing (EUC)*, New York, NY, USA, pp. 367–372 (2019). https://doi.org/10.1109/CSE/EUC.2019.00076
18. M. Singh and S. Kim, Security analysis of intelligent vehicles: Challenges and scope, in *2017 International SoC Design Conference (ISOCC)*, pp. 13–14 (2017). https://doi.org/10.1109/ISOCC.2017.8368805
19. A. Eskandarian, Introduction to Intelligent Vehicles, in *Handbook of Intelligent Vehicles* ed. by A. Eskandarian, (Springer, London 2012). https://doi.org/10.1007/978-0-85729-085-4_1
20. I. Singh, S. Lee, Self-adaptive requirements for intelligent transportation system: A case study, 2017 International Conference on Information and Communication Technology Convergence (ICTC), pp. 520–526 (2017). https://doi.org/10.1109/ICTC.2017.8191032
21. M. Singh, Requirement engineering for intelligent vehicles at safety perspective, in *EAI Endorsed Transactions on Smart Cities*, **2**(6), (2017), ISSN No. 2518-3893. https://doi.org/10.4108/eai.20-12-2017.153496

Chapter 14
Blockchain Technology in Vehicular Communication Networks

Madhusudan Singh and Hasan Tinmaz

Abstract In present scenario, Blockchain is a rapidly emerging technology that can provide highly secure decentralized distributed data between two nodes in a very large network. It also keeps all the broadcasted data in ledger from beginning to current time without exposing any personal information. We believe that Blockchain has capability to remove current VCN issues such as: Centralization, Data loss, Data spoofing, Untrusted message broadcast, DDoS issues, and Privacy. This paper we have introduced blockchain technology to provide a decentralized distributed secure and safe vehicle communications environment. In our proposed model, vehicles can communicate peer to peer node without exposing any private information and other party will also trust the broadcasted message and response accordingly. We have verify the blockchain technology for vehicular communication with simulation for intersection use case.

Keywords Blockchain technology · Vehicular communication · Secure trust

14.1 Introduction

Vehicular Communication Network (VCN) has been an area of dynamic research in recent years. However, most of the research focuses on safety, and IT security, which mostly concern about centralized security, and key management. Our current vehicular communication environment has a lot of loopholes such as, centralized Transportation System, Untrusted message broadcast, easy-to-hack, Lack of information about DDoS, Lack of Privacy, Data loss and Spoofing. As we know our vehicles are going to be smarter day by day and also design goal is to make them smaller as well as augmenting with more technologies such as sensors, ECU, CAN

M. Singh · H. Tinmaz (✉)
School of Technology Studies, Endicott College of International Studies, Woosong University, Daejeon, South Korea
e-mail: htinmaz@endicott.ac.kr

M. Singh
e-mail: msingh@wsu.ac.kr

© The Author(s), under exclusive license to Springer Nature Singapore Pte Ltd. 2021   213
M. Singh, *Information Security of Intelligent Vehicles Communication*,
Studies in Computational Intelligence 978,
https://doi.org/10.1007/978-981-16-2217-5_14

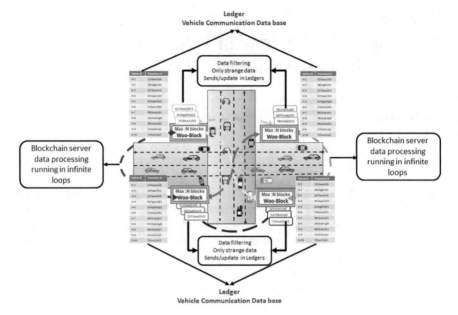

Fig. 14.1 Proposed blockchain based secure trust environment model for vehicular communication network (BSTEM-VCN)

etc. Coming few years are going to bring a revolutionary change in vehicular communication network. To improve VCN from its current state to the futuristic vehicles communication, we have proposed Blockchain based secure trust environment model for VCN (BSTEM-VCN) as represents in Fig. 14.1.

Vehicular Communication Networks (VCN) is an ongoing emerging and appealing communication technology for our society to provide better and safer vehicle network and their services. It includes other specific types of communication as V2I (Vehicle-to-Infrastructure), V2V (Vehicle-to-Vehicle), V2P (Vehicle-to-Pedestrian), V2D (Vehicle-to-Device) and V2C Vehicle-to-Cloud) [1].

Figure 14.2 shows the VCN Model, which includes V2V, V2I, V2X communication environment. The VCN is unified concept for cyber physical system which includes, a set of existing and emerging cyber world (wireless technologies, sensors, control system), Physical entities (vehicles, Infrastructure etc.), which are combined and connected with a systematic way. The research results from these application areas greatly contribute to the development, implementation, and emergence of VCN.

In VCN vehicles are equipped with Internet access, and with a wirelessly connected networks. This allows the vehicles to share there requirements such as services, message transmission, etc. through communication channels with other vehicles or devices both inside of vehicle as well as outside the vehicle communication infrastructure such as transportation system, service centers or other vehicles etc. The Intelligent Transportation System (ITS) provides safe, efficient flow of traffic, services as well as traveler information services. ITS is provide a support system

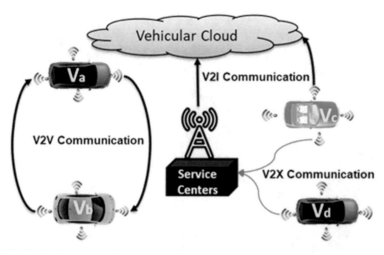

Fig. 14.2 Need of research vehicular communication networks

where vehicles can be share or exchange information among vehicles and infrastructures through the communication protocols such as vehicle-to-vehicle (V2V),
vehicle-to-(roadside) infrastructure (V2I), vehicle-to-network (V2N), and vehicle-
to-pedestrian (V2P) or vehicle-to-everything (V2X) communication [2]. In VCN
networks, vehicles are connected with almost everything. As much as vehicles
are in communication environment, thus the Cybersecurity issues increase in the
VCN. In the VCN environment protection of vehicle personal information, message
transmission, data integrity and security of vehicles as well as safety related functions become big challenges. To solve above mentioned challenges, we have introduce Blockchain Technology in vehicular communication network environment [3].
The our proposed method blockchain technology can be provide privacy secure
decentralized distributed and trusted environment for vehicular communication
networks.

This article has organized as follows: Sect. 14.2 has provide the background
study about vehicular communication and blockchain technology, Sect. 14.3 has
explain the proposed Blockchain based secure trusted environment model for vehicular communication networks. Section 14.4 shown the implementation of proposed
idea in Intersection use case and in Sect. 14.5, we have conclude the our article.

14.2   Background Study

Vehicular Communication Networks are evolving rapidly as a communication environment where vehicles talk with everything. Whenever any vehicle is connected to
any communications medium, it is exposed to a risk from malicious cyber attacks. The
automotive industries are introducing the most computerized features and networked

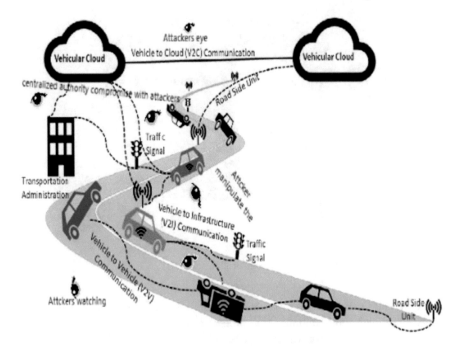

Fig. 14.3 Security challenges in vehicular communication networks

enabled vehicles so as to whenever it can talk to anything, it opens new doors for hacking. Like the other information and communication industries that are often targets of data breaches, the vehicular communication networks [4] must consider the many potential attack surfaces through which hackers may try to break into a network. The attack surface of VCN is broad and may go way beyond the vulnerability of the vehicle itself. The security threat of vehicular communication networks can be explained in 4 ways—illegal remotely access of vehicle within networks, attack on vehicles through communication channels, passive attack by tampering sensors data, try to access in-vehicle network connectivity architecture. In Fig. 14.3, we can see the potential issues of cybersecurity in VCN networks.

14.2.1 Blockchain Technology: An Overview

Blockchain is a disruptive technology that emerged in late 2008, in the midst of the global financial crisis. It is comprised of unchangeable, digitally recorded data in packages called blocks. These digitally recorded "blocks" of data is stored in a linear chain. Each block in the chain contains data that is cryptographically hashed.

Blockchain Technology   **Automotive Technology**

Fig. 14.4 Relationship between blockchain technology and vehicular communication networks

- **Understanding the Basics of Blockchain Technology**

A digital distributed ledger is a consensus of replicated, shared, and synchronized digital data geographically spread across multiple sites. One form of distributed ledger design is the blockchain system, which can be either public or private. A Blockchain is a type of a digital distributed ledger [5].

As stated before, to better understand blockchain technology we must consider the distributed ledger technology is a "parent" technology of the blockchain. As a result, all blockchain technologies are indeed a type of distributed ledger but not all distributed ledgers are blockchains! As far as the architecture behind blockchain technology it is a peer-to-peer network without the interference of a third party [6].

- **Relationship Between Blockchain Technology and Vehicular Communica-tion Networks?**

What are those characteristics of Blockchain Technology that lend themselves to vehicular Communication Networks and their everyday use? as shown in Fig. 14.4.

The pinnacle accomplishment of the merging of the two industries is the development of safe autonomous vehicles that utilize the security characteristics offered by blockchain. This promotes safety which is the paramount prerequisite in the automotive industry.

Let compare, the traditional security offered by the information technology industry as compared to the new blockchain solution. Traditional security methods are incapable to provide secure communication for vehicles. First traditional protection functions on a centralized architecture that does not apply itself to the decentralized vehicular reality, for example it does not offer peer to peer communication. Second, it requires a lot of resources in order to offer adequate protection, such resources are encryption and decryption enabled hardware. Third, it is very hard for traditional systems to protect data that is mobile—on the go like in the case of vehicles that are moving. Fourth, they are unable to manage and protect large amounts of data being exchanged at the same time. Fifth they do not have access to the history of each vehicle and therefore there is no verification and Sixth there is no distributed network. On the other hand Blockchain Technology protection offers and that is why we need blockchain technology.

When comparing Traditional versus Blockchain Protection there are numerous characteristics that we must explain in order to understand the true value that blockchain technology has to offer.

Blockchain Technology Protection is by definition a decentralized architecture that mimics the vehicular environment. Second, it is low on resources since one blockchain server can manage the entire system. Third, it is able to protect data that is mobile because it is decentralized and resources are distributed throughout the vehicle communication environment. Fourth, it is able to handle large amounts of data at the same time because the data is stored in the blockchain server which is distributed and not centralized. Fifth, it has access to vehicle history because of peer to peer real time communication and that all the data is stored on the distributed block chain server. Finally, there is a distributed network since every node has it's own database. There, on those databases the same network data is shared in real time between the nodes [7].

- **The Relevant Blockchain Characteristics That are Important to Vehicles Safety in Vehicular Communication Environment?**

In the blockchain paradigm, all parties must agree to network verified transaction, therefore, we have Consensus. The system is what is called an append-only system of record that is shared across a business network, therefore, we have a Shared Ledger. All the business terms are embedded in a transaction database and executed with transactions, therefore we have what is called Shared Contract. Last, the blockhain paradigm ensures secure, authenticated and verifiable transactions, therefore we have Cryptography. We have explained in details as below

- **Consensus**

Consensus means that all parties agree to network verified transaction. This characteristic promotes Trust ability Liability, and Authentication/Authorization as shown in Fig. 14.5.

Fig. 14.5 The characteristic of consensus

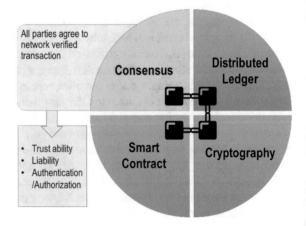

Fig. 14.6 Features of smart
contract

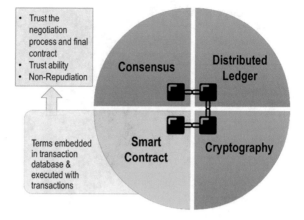

– **Smart Contract**

Having shared contract means that all terms are embedded in a transaction database
and executed with transactions. This ensures a Trust in the negotiation process and
the final contract. Thanks to this we have Trust ability and Non-Repudiation which
is the assurance that someone cannot deny something. Figure 14.6 has represents the
features of Smart Contract.

– **Distributed Ledger**

Having a distributed Ledger, means that the system will store all communication
information data stored into the ledger. This is what is called an append-only system
of record, because it is stored and distributed across a network. Therefore, we have
Availability, Retain ability and Traceability. In Fig. 14.7, we have represents the
explanation of Distributed Ledger.

Fig. 14.7 Distributed ledger
overview

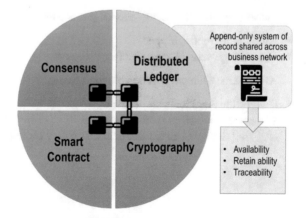

Fig. 14.8 Cryptography
features

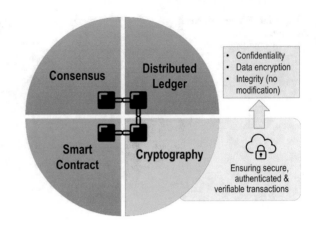

– **Cryptography**

Having Cryptography means secure, authenticated and verifiable transactions which translates into confidentiality, data encryption and Integrity since no modification can be made. In autonomous vehicles we achieve this by using the Hash ID. Cryptography features represents in Fig. 14.8.

This is what the blockchain network looks like in the vehicular communication environment. Vehicles will be able to communicate from vehicle to vehicle all information with privacy [8]. The vehicles can communicate to the vehicular cloud and receive information about services. All vehicles share information with family or friends within the Wi Fi infrastructure. All vehicular information is stored on the Blockchain server. This is happening in real time and the data stored can be accessed any time. Figure 14.9 has represents the blockchain based vehicular communication networks architecture.

– **Blockchain based communication networks provides following benefits**:

The motivation for Number of Passengers traveling

- Amount of Goods shipped
- The average vehicle speed by far.

    At the same time we will be able to decrease:

- Congestion Accidents and Crashes
- The overall industry Carbon Footprint
- Utilizing the Same Basic Road and Highway Infrastructure.

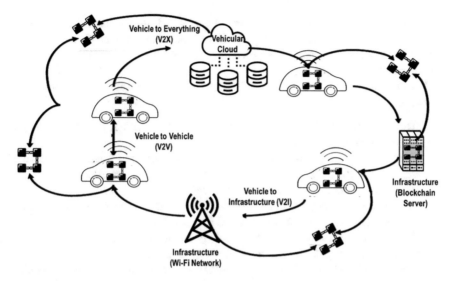

Fig. 14.9   Blockchain based vehicular communication networks architecture

14.3   Proposed Blockchain Based Secure Trust Environment Model for Vehicular Communication Networks

This research aims to develop a Blockchain based secure trust environment (BSTEM), where vehicles communicate, share and manage information between anything without sharing any personal information. In our proposal, we will develop a new Blockchain framework for vehicular communication networks and also, we define new consensus and cryptography algorithm with data management according to VCN environment in which vehicles can seamlessly, fearlessly and trustfully share the data. Our research on BSTEM mechanism will be used to reduce the cybersecurity issues and build a trust in VCN.

This work is different from the schemes proposed in the literature because it focuses on real time vehicular communication networks and energy efficient communication between tasks. In addition, it aims to address task failure by adopting a failure avoidance mechanism and to make effective resource allocation decisions, a hybrid architecture will be used. This scheme will be based on a cross layer design approach in which network information will be shared with resource management system for effective decision making. Our proposed research will develop new Blockchain algorithm for VCN with mobility, data management and security enabled features within low power consumption. It will manage the complete process of Blockchain technology under small computation power enable vehicles. We will provide the Proof of work and Consensus algorithm according to real time vehicles features. Our research will provide a transmission power control mechanism which will be used to reduce transmission power energy consumption and increase concurrent transmissions in the network. The new system will consider the multi-radio nodes which are used to increase network capacity and support large data management and transfers (Ledger). The Fig. 14.10 has shown the as example of complete process of blockchain based secure trust environment mode (BSTEM) in vehicular communications networks.

Let us examine how the process of BSTEM works in Vehicular Communication Networks:

1. The blockchain enabled vehicle on the left acts as the sender. The sender creates the transaction symbolized here as a block.
2. Transaction distributed and validated via cryptographic hashing. After the verification, the information is stored on the blockchain servers.
3. The same information is mirrored on all servers.
4. All this is part of the Vehicular Cloud in which Data transmission is committed to blockchain and miners are rewarded by the senders.
5. The vehicular cloud is a hybrid technology that uses vehicle resources, such as computing, data storage, and internet decision-making.
6. The vehicular cloud is a hybrid technology that uses vehicle resources, such as computing, data storage, and internet decision-making.
7. A "miner" is a blockchain enabled vehicle that can validate the data transmission in the network (Fig. 14.10).
8. The receiver is confident that the information is accurate because it is coming from blockchain server.

Note: The reward system is important because it builds trust among unrelated vehicles that are unrelated in any way yet now become trusted and this is done without forfeiting privacy. That is why messages are valuable only when they have been verified and validated. This is why a lot of time and energy goes into data mining which actually is data verification. The system maximizes trust without compromising privacy.

For reward purpose, we have introduced Vehicle Trust Point (VTP) which consists of random alphanumeric numbers that represent a crypto ID [9]. A crypto ID is used to establish trust in the entire network. Among the vehicles, among the vehicles and the infrastructure, etc. This crypto ID is different every time it is transmitted

Fig. 14.10 Process of blockchain based secure trust environment model (BSTEM) in vehicular communication networks

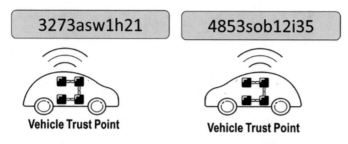

Fig. 14.11 Representation of Vehicle Trust Point (VTP)

and is used as reward to all those vehicles that are miners. The vehicle with the highest number of VTPs is the most trusted vehicle in the network. The more data mining a vehicle conducts in the network the more rewarded it is by other vehicles as represented in Fig. 14.11.

14.4   Simulation for Intersection Scenario

For the simulation of the intersection scenario, pygame library of Python was used. The first step of the simulation is placing four vehicles in four different directions of the simulation. Initially the coordinates of each vehicle are near the edge of the frame. The intersection can be used by only one vehicle at a time. Random velocities are allotted to the vehicles at the beginning. There is also a display associated with each vehicle which tells how many VTPs it is currently having. When the simulation receives the start packet, the vehicle starts moving and when it reaches the intersection, a flag is raised and the timestamp is sent to the blockchain logic. Then this transaction is mined and it is decided which car is to be moved first. I restriction is then set on all the other vehicles to disallow them to move until this vehicle crosses the intersection. The one who mined the timestamp transaction block is awarded some VTP.

The simulation generated random velocities and also we could give random starting positions to vehicles which enables us to test various scenarios. In our simulation IV represents Intelligent Vehicles.

- **Step by step demonstration of the blockchain**

1.   The initial position, none of the vehicles have started yet as none of the start blocks has been received yet by the simulation as shown in Fig. 14.12.
2.   After some time, each vehicle sends its start block as they start moving towards the intersection as shown in below Fig. 14.13 .
3.   Vehicle on the left reaches the intersection first. Hence it broadcasts the "reaches intersection" packet to all the vehicles, represents in Fig. 14.14.
4.   As we can see, the vehicle on the left was allowed to move first and crossed the intersection. The next one to come in was IV 4 (down), hence it is allowed to

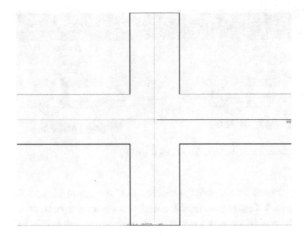

Fig. 14.12 Initial position

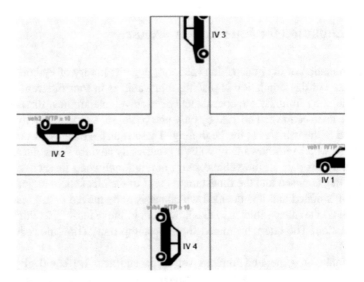

Fig. 14.13 Vehicles start moving

cross the intersection. IV I (Vehicle on the right) has successfully mined two transactions. Hence it has 12 VTPs and the VTPs of IV 4 and IV I are reduced to 9 each. We can see in Fig. 14.15.

5.   Since IV I reached last, it has to wait for all other vehicles to cross the intersection. As it also mined blocks, it is awarded VTP represents in Fig. 14.16.

In order to find out how good the developed ITS Blockchain module was, it was necessary to make a simulation testbed. Also, a Communication Module had to be built for the simulation purpose. A straightforward Communication Module based

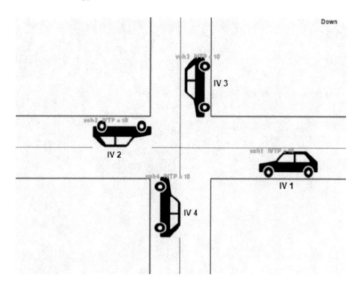

Fig. 14.14   First vehicle reaches intersection

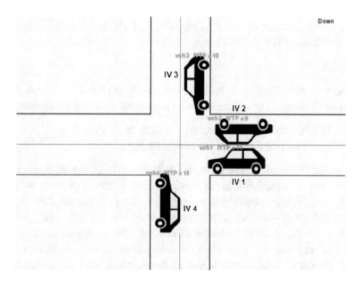

Fig. 14.15   Vehicle moves after consensus

on Client–Server model was built as its low latency would help to measure the performance of the blockchain system without it being influenced by the performance of the network. Also, it was decided that the simulation be run on a local Wireless Area Network for low latency. Using such an environment, the ITS Blockchain module was tested using a five-computer setup. We have tested our proposed method

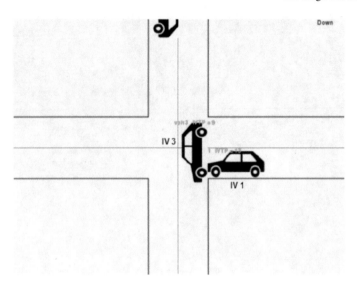

Fig. 14.16 All vehicles move

3 difficulty level (number of zeros in beginning of the HASH which take average 1.03083 s. for mining for our 100 trials.

- **Discussion**

Tests in single-branch mode was done to measure the performance of the underlying blockchain technology. The questions to be answered were the following:

– Does the ITS Blockchain network require time to stabilize?
– Does the performance increase/decrease with time?

Owing to the dynamic nature of traffic, it is key that all the state changes in the network be processed (added into the blockchain or rejected) as soon as it is made.

From Fig. 14.17, it can be seen that the graphs are irregular till around t = 400 s, and is fairly steady afterwards. It means that once the network has stabilized, the mining rate is constant for a given network load. The performance remains same irrespective of time elapsed, given that the network load does not change.

Another important observation is that as the network load increases, the mining rate decreases (lesser and lesser number of blocks are mined per unit time). This poses us a question—if the mining rate is lesser for higher network load, does it mean that the number of state changes verified in unit time is also low? The answer for this can be understood from the following graph.

From Fig. 14.18, we see that the state change verification rate increases in proportion with the increase in network load. This implies that the fact that mining rate is low for higher network load does not affect the number of state changes verified.

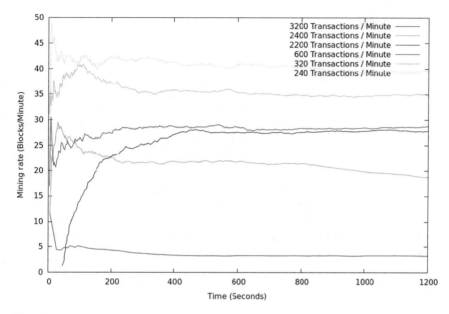

Fig. 14.17 Mining rate versus time (for various network loads)

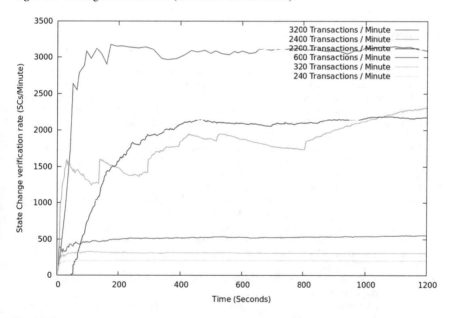

Fig. 14.18 State changes verification rate versus time (for various network loads)

14.5 Conclusion

Our proposed Blockchain based Secure Trust Environment Model for Vehicular Communication Networks (B-STEM: VCN) has provide secure vehicular communication with trusted service information. It has protected to vehicles to transmits or broadcast fake data in networks and also obtain history of any vehicles, connected in the networks. The main challenge faced by using blockchain to authenticate and validate inter-vehicular (V2V) and vehicle to infrastructure (V2I) communication is the dynamic nature of the traffic. Hence, the qualities an ideal vehicle blockchain should have are two fold—it should be fairly real time, and it should be able to handle a huge amount of message traffic without getting choked. We have try to find the feasibility of blockchain technology for vehicular communication network. As based on our simulation, we can say that blockchain technology can be a possible good solution for future high-tech vehicles such as self driving car. As future work, we will try to implement in multiple and more complicated USE CASE Scenario.

Acknowledgements This research was funded by Woosong University Academic Research in 2021.

References

1. I.A. Abbasi, K.A. Shahid, A review of vehicle to vehicle communication protocols for VANETs in the urban environment. Future Internet **10**, 14 (2018)
2. G.-U. Rehman, A. Ghani, S. Muhammad, M. Singh, D. Singh, Selfishness in vehicular delay-tolerant networks: a review. Sensors **20**, 3000 (2020). https://doi.org/10.3390/s20103000
3. M.S. Dorri, S.S. Kanhere, R. Jurdak, Blockchain: a distributed solution to automotive security and privacy. IEEE Commun. Mag. **55**(12), 119–125 (2017)
4. M. Singh, Blockchain technology for data management in Industry 4.0, in *Blockchain Technology for Industry 4.0. Blockchain Technologies*, ed. by R. Rosa Righi, A. Alberti, M. Singh (Springer, Singapore, 2020). https://doi.org/10.1007/978-981-15-1137-0_3
5. M.J. Diván, M. Singh, The impact of the measurement process in intelligent system of data gathering strategies, in *Intelligent Human Computer Interaction. IHCI 2020*, ed. by M. Singh, D.K. Kang, J.H. Lee, U.S. Tiwary, D. Singh, W.Y. Chung. Lecture Notes in Computer Science, vol. 12615 (Springer, Cham, 2021). https://doi.org/10.1007/978-3-030-68449-5_43
6. L. Zhang, M. Luo, J. Li, M.H. Au, K.-K. Raymond Choo, T. Chen et al., Blockchain based secure data sharing system for Internet of vehicles: a position paper. Veh. Commun. **16**, 85–93 (2019). ISSN 2214-2096
7. J.R. Reagan, M. Singh, Automotive evolution, in *Management 4.0. Blockchain Technologies* (Springer, Singapore, 2020). https://doi.org/10.1007/978-981-15-6751-3_2
8. S. Zeadally, J. Guerrero, J. Contreras, A tutorial survey on vehicle-to-vehicle communications. Telecommun. Syst. **73**, 469–489 (2020)
9. M. Singh, Tri-blockchain based intelligent vehicular networks, in *IEEE INFOCOM 2020—IEEE Conference on Computer Communications Workshops (INFOCOM WKSHPS)*, Toronto, 2020, pp. 860–864. https://doi.org/10.1109/INFOCOMWKSHPS50562.2020.9162692

Chapter 15
Adaptive Proof of Driving Consensus for Intelligent Vehicle Communication

Madhusudan Singh and Iftekhar Salam

Abstract Energy and cost efficiency is a notable drawback in Blockchain Technology as well as in traditional consensus schemes and Blockchain does not have any standard consensus protocol for resource constraint applications such as vehicular communication environment. In this paper, we propose a novel Adaptive Proof of Driving (APoD) consensus algorithm for Blockchain based Vehicular Communication Networks (VCN), especially for resource constraint environments. Our developed algorithm provides secure, trusted and reliable communication environment for vehicles based on reputation. The main contribution of our paper is to improve the power consumption, miner congestion, and cost efficiency of blockchain system for vehicular networks. We evaluate, our proposed approach with the help of congestion less intersection use case for adaptive and secure distributed consensus.

Keywords Blockchain technology · Consensus vehicular communication · Intelligent transportation system

15.1 Introduction

Blockchain technology has the capability to contribute to several fields. Blockchain can contour the eHealth record process and contribute better visibility and efficiency over the supply chain to provision eminent value to the customers and trading relations. Moreover, Blockchain can trace possession of real estate and alter the mode of sharing data and forbidding fraudulence. Blockchain can also revolutionize the Internet of Things and vehicular communication [1].

M. Singh (✉)
School of Technology Studies, Endicott College of International Studies, Woosong University, Daejeon, South Korea
e-mail: msingh@wsu.ac.kr

I. Salam
School of Electrical and Computer Engineering, Xiamen University Malaysia, 43900 Sepang, Selangor, Malaysia
e-mail: iftekhar.salam@xmu.edu.my

The vehicle is experiencing revolutionary growth in research and industry such as intelligent vehicles (IV), vehicle communications, self-driving vehicles, but it still suffers from many security vulnerabilities. Traditional security methods are incapable to provide secure vehicular communication. The major issues in vehicular communication are trust, data accuracy and reliability of communication data in the communication channel [2]. Blockchain technology works for the cryptocurrency, Bit-coin, to develop trust and reliability in peer-to-peer networks which have alike topologies to Vehicle Communication Networks (VCN). But Blockchain doesn't have any standard consensus protocol for resource constraint applications such as vehicular communication environment. In this paper, we propose a novel Adaptive Proof of Driving (APoD) consensus algorithm for Blockchain based Vehicular Communication Networks (VCN), especially for resource constraint environments. Our developed algorithm will provide secure, trusted and reliable communication environment for vehicle based on reputation. The aim of our paper is to improve the power consumption, miner congestion, and cost efficiency of blockchain system for vehicular networks. We evaluate our proposed approach with the help of congestion less intersection use case for adaptive and secure distributed consensus. We have depicted in the Fig. 15.1 is current vehicle communication environment.

The remainder of the work is outlined as follows, Sect. 15.2 has discussed a review of related work, and further, we have explained our proposed consensus algorithm for intelligent vehicles communication environment in Sect. 15.3. We have discussed an intelligent vehicle communication use case model in Sect. 15.4. At last we have concluded our article in Sect. 15.5.

15.2 Related Work

It is challenging to build a secure distributed consensus scheme but there are tremendous benefits for several applications. Bitcoin and other similar cryptocurrencies used in the financial technology are familiar. Bitcoin has several insufficiencies and several altcoin [3] schemes have been proposed to make them better but still not all deficiencies are sufficiently covered. All present cryptocurrency approaches concentrates on payment transaction and maintaining fairness is critical in such transactions as payments are done in exchange of service or goods and there is no identity reliability of the participants and also we cannot practically resolve any subsequent dispute. Another limitation is link ability, making personally identifying the Bitcoin transaction [4]. Zero cash [5] is an alternate which provision payer and payee unlink ability earned as altcoin strategies [6]. Other applications of distributed consensus other than Fin tech are driverless vehicles [7], autonomous robots, supply chain management and sharing information in the Internet of Things (IoT).

There are different requirements for every application and it is critical to decide a distributed consensus which is compatible, suitable and fulfill the requirements for that application. Energy efficiency is a critical requirement and blockchain based on

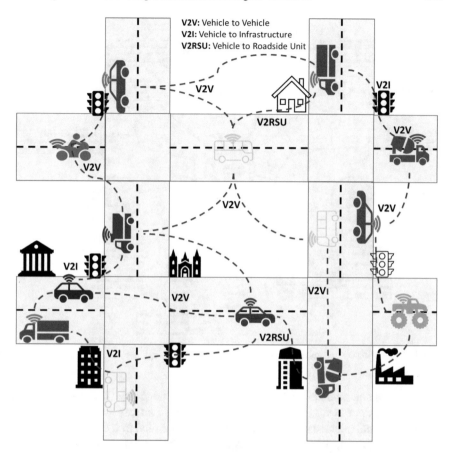

Fig. 15.1 Current vehicle communication environment [1]

"proof of work" consensus scheme like Bitcoin are not favorable in terms of usage of energy.

The traditional Byzantine fault tolerant scheme also consume a lot of energy as every participant need to send and receive the broadcasted message. We aim to develop a consensus schemes which are energy efficient. Figure 15.2 shows several existing consensus algorithms and their work.

- ***Challenges of Existing Blockchain Consensus***

As of now, there is no standard Consensus scheme fully compatible for Vehicular Communication Network (VCN). However, researchers are using traditional consensus schemes (CheapBFT [8], MinBFT/MinZyzzyva [9], and Proof of Work (PoW) [2]) in VCN. There are various limitations of consensus schemes in Distributed Computing such as safety (all valid participant's agreement on consistent value) and liveness (all valid participants finally determine a value). According to Bracha and Toueg [8] if more than f participants are Byzantine then consensus in a loosely

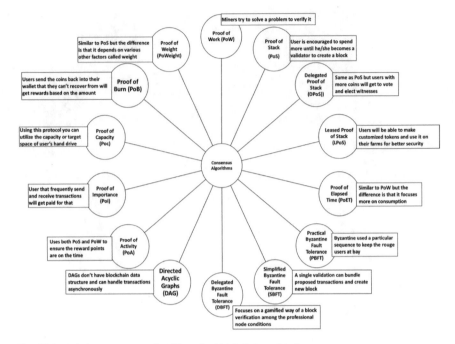

Fig. 15.2 Existing consensus algorithms for blockchain technology

synchronous communication model is not feasible among $n = 3f + 1$ participants. Likewise, Brewer et al. [9]'s "CAP theorem" states that consistency, availability and partition-tolerance cannot be assured at the same time in a distributed system. Classical PBFT [10] consensus schemes are established on state machine replication and are becoming obsolete due to their huge cost of computation, bandwidth and expected number of autonomous servers [11].

15.3 Proposed Adaptive Proof of Driving (APoD) Methodology

We have developed a decentralized distributed reputation based Vehicular Communication Networks (VCN) that use Blockchain technology to build trust among vehicles. The proposed Blockchain technology for vehicular communication is based on novel consensus algorithm, Adaptive Proof of Driving (APoD) which, increases performance and throughput by reducing miner congestion. Our proposed algorithm provides secure reliable scalable consensus mechanism for vehicular network, by combining two schemes, First Come First Serve (FCFS) and Priority (Reputation) scheme as shown in step 1 and step 2 in Fig. 15.2. The FCFS scheme maintains the timestamp list using the blockchain server and priority scheme gives preference to lower timestamp and reputed Vehicles. To validate the reputed vehicles, we have introduced Vehicle Reputation Point (VRP) which is an issued unique crypto ID for each vehicle and the same VRP are used to enable the flow of reputation points, which act as a reputation value for vehicle to get involved in the information exchange between vehicles. For the data management of the VRP, we are using blockchain technology in the intelligent transportation system (ITS), which stores all VRP details of every vehicle and is accessed ubiquitously by vehicles [12].

- *Adaptive Proof of Driving (APoD) Consensus Algorithm*

We propose an Adaptive Proof of Driving (APoD) consensus algorithm which uses two schemes, First Come First Serve (FCFS) and Priority to drop-off the congestion of miners to increase the performance. Figure 15.3 shows our proposed Consensus approach. The two steps step 1 and step 2 are defined below.

- *First Come and First Serve (FCFS)*: First Come and First Scheme uses timestamp t_s to reduce the miner congestion in the vehicular communication network. If timestamp t_i of a vehicle V_i is less than the timestamp t_j of a vehicle V_j, then vehicle V_i is added in the miner list as it has lower timestamp.
- *Priority Scheme*: The blockchain server maintains the miner list of first come messages based on the Vehicular Reputation Point (*VRP*) of each vehicle. The

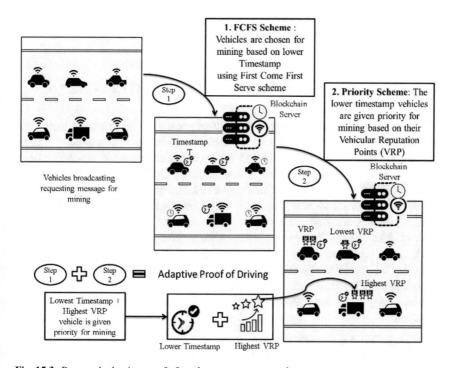

Fig. 15.3 Proposed adaptive proof of work consensus approach

priority scheme uses the Vehicular Reputation Points (VRP) to decide the priority given to vehicles in the Miner list. The higher priority vehicles are trusted and given incentives in terms of VRP. If VRP of a vehicle V_i is greater than the VRP of a vehicle V_j, then vehicle $V_i's$ broadcasted message is trusted and is awarded some Reward Points.

Below is the pseudo code of APoD Consensus scheme:

Pseudocode of APoD Consensus Scheme

$FCFS(V(t_s))$

1. *For timestamp instant t_s*
2. *If $(V_i(t_i)) < (V_j(t_j))$*
3. *$Miner_{list} \leftarrow V_i$*
4. *return $Miner_{list}$*
    $Priority(Miner_{list}(VRP))$
5. *For Vehicular Reputation Points VRP*
6. *If $(VRP[V_i] > VRP[V_j])$*
7. *$VRP[V_i] \leftarrow VRP[V_i] + Reward\ Points$*
8. *return $VRP[V_i]$*
9. *else $VRP[V_j] \leftarrow VRP[V_j] + Reward\ Points$*
10. *return $VRP[V_j]$*

Our proposed APoD algorithm is energy efficient, cost efficient and builds trust and reliability in the blockchain network.

15.4   Intelligent Vehicle Communication Use Case Scenario

We discuss the intersection scenario for, utilizing cost and energy and building trust and reliability in the blockchain based vehicular communication network. In an intersection scenario as shown in Fig. 15.4, all vehicles are coming from all directions at the intersection and all vehicles cannot cross the intersection simultaneously as deadlock situation will happen. Therefore, vehicles must cross the intersection one at a time to avoid deadlock situation. Now, the problem arises as to which vehicle will cross the intersection first and in what order so that all vehicles can cross and there is no deadlock situation. To solve this problem, we applied our proposed Adaptive Proof of Driving (APoD) consensus algorithm which uses two schemes, First Come First Serve (FCFS) and Priority to drop-off the congestion of vehicles, deciding which vehicle to cross the intersection first thereby utilizing energy, cost and building trust and reliability in the blockchain network. The description of each of the scheme is as follows:

First Come First Serve (FCFS) Scheme: In this scheme the vehicle which arrives first at the intersection, is stored in the miner list which is later used by the priority

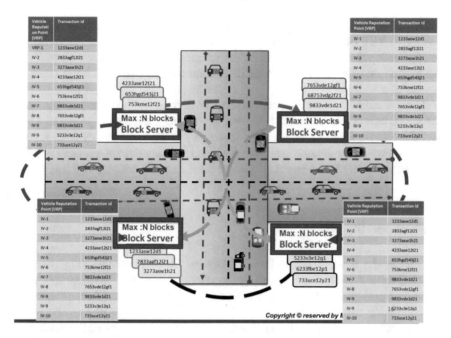

Fig. 15.4 Overall intelligent vehicle communication based on adaptive proof of driving consensus algorithm

scheme to determine the priority to mine based on the vehicle's Vehicle Reputation Point (VRP) to find the order of vehicles to cross the intersection.

For example: Assume four vehicles V_1, V_2, V_3, *and* V_4 arrives at the intersection at 1.65, 1.63, 1.66, 1.61 s respectively, then the vehicle whose arrival time is minimum (arrives first at intersection will be eligible as miner). So, V_4 with minimum arrival time of 1.61 s will be the first miner added in the miner list. Thereafter, V_2, V_1 and V_3 will be given preference for mining in ascending order of their arrival time [12].

Priority Scheme: In this scheme, each vehicle in the miner list is assigned a priority based on the Vehicle Reputation point (VRP) it owns as shown in the distributed ledger of each blockchain server in Fig. 15.4. The amount of VRP indicates the trust and reliability the vehicle has. The greater the VRP, the greater is the vehicle's trust and highest will be the vehicle's priority. The highest priority vehicle with the lowest arrival time at the intersection will first cross the intersection and the lowest priority vehicle with highest arrival time at the intersection will cross the intersection at the last. The miner vehicle will generate the table accordingly.

Vehicles with same VRP or same priority will follow the FCFS scheme to decide the order in which they will cross the intersection.

For example: The vehicles V_1, V_2, V_3 and V_4 have 10, 15, 20, and 35 VRP respectively and the miner list ascending order is { V_4, V_2, V_1, *and* V_3} then the vehicle V_4, having lowest arrival time of 1.61 s and greatest VRP of 35 will be considered the most trusted and reliable vehicle in the network of the four vehicles. Similarly, vehicle V_1 having 10 VRP (minimum in the network) will be considered the least trusted and least reliable vehicle in the network of the vehicles. Therefore, vehicle V_4 will have the highest priority to do the mining to find out the order of vehicles to cross the intersection.

Note: In same VRP and Time stamp then we will consider 2nd significant figures of the arrival time of the vehicles at the intersection. After applying our proposed APoD algorithm to the intersection scenario shown in Fig. 15.4, the vehicular communication becomes seamless avoiding deadlock and utilizing energy, cost and building trust and reliability in the blockchain based vehicular network.

15.5 Conclusion

We proposed a novel Adaptive Proof of Driving (APoD) consensus algorithm for Blockchain based Vehicular Communication Networks (VCN). Our proposed APoD algorithm provides secure reliable consensus mechanism for vehicular network, by combining two schemes, First Come First Serve (FCFS) and Priority (Reputation) scheme. The FCFS scheme maintains the timestamp list using the blockchain server and priority scheme gives preference to lower timestamp and reputed Vehicles thereby improving the power consumption, miner congestion, and cost efficiency of blockchain system for vehicular networks. To validate the reputed vehicles, we have introduced Vehicle Reputation Point (VRP) which is an issued unique crypto ID for each vehicle and the same VRP are used to enable the flow of reputation points, which act as a reputation value for vehicle to get involved in the information exchange between vehicles. In future, we will implement our consensus algorithm for Blockchain-based secure decentralized reliable vehicular communication networks through simulations and real-life experiments.

Acknowledgements This research was funded by Woosong University Academic Research in 2021.

References

1. M. Singh, S. Kim, Crypto trust point (cTp) for secure data sharing among intelligent vehicles, in *The 2018 International Conference on Electronics, Information and Communication (ICEIC 2018)*, Sheraton Waikiki Hotel, Honolulu, Hawaii, USA, 24–27 Jan 2018. https://ieeexplore.ieee.org/document/8330663/
2. B. Leiding, P. Memarmoshrefi, D. Hogrefe, Self-managed and blockchain-based vehicular ad-hoc networks, in *Proceedings of the ACM International Joint Conference on Pervasive and Ubiquitous Computing*, 2016, pp. 137–140. https://dl.acm.org/citation.cfm?id=2971409
3. S. Kim, Blockchain for a trust network among intelligent vehicles. Adv. Comput. **111** (2018). ISSN 0065-2456. https://www.sciencedirect.com/science/article/pii/S0065245818300238

4. G.S. Veronese et al., Efficient byzantine fault-tolerance. IEEE Trans. Comput. **62**(1), 16–30 (2013). https://ieeexplore.ieee.org/document/6081855/?arnumber=6081855
5. R. Kapitza et al., CheapBFT: resource-efficient byzantine fault tolerance, in *EuroSys* (ACM, 2012). https://doi.acm.org/10.1145/2168836.2168866
6. M. Vukolić, The quest for scalable blockchain fabric: proof-of-work vs. BFT replication, in *International Workshop on Open Problems in Network Security* (Springer, 2015). https://link.springer.com/chapter/10.1007/978-3-319-39028-4_9
7. L. Luu et al., Demystifying incentives in the consensus computer, in *CCS* (ACM, 2015). https://doi.acm.org/10.1145/2810103.2813659
8. Bracha, S. Toueg, Asynchronous consensus and broadcast protocols. J. ACM **32**(4), 824–840 (1985). https://dl.acm.org/citation.cfm?id=214134
9. E.A. Brewer, Towards robust distributed systems (abstract), in *PODC* (ACM, 2000). https://doi.acm.org/10.1145/343477.343502
10. M. Castro, B. Liskov, Practical byzantine fault tolerance, in *OSDI* (USENIX Association, 1999). https://dl.acm.org/citation.cfm?id=296806.296824
11. V. Buterin, A next-generation smart contract and decentralized application platform (2014). https://github.com/ethereum/wiki/wiki/White-Paper
12. M. Singh, S. Kim, Branch based blockchain technology in intelligent vehicle. Comput. Netw. (2018). ISSN 1389-1286. https://doi.org/10.1016/j.comnet.2018.08.016

Appendix A
Scenarios

Red text = actor
Green highlight = goal

1. Starting the vehicle

    1. No problems

        The driver enters the car and starts the Car. The On Board Diagnostic diagnoses the vehicle; it checks the working of different sensors like temperature sensor, pressure sensor, proximity sensor and other sensors. It also monitors emission control, mileage, and speed and fuel status.

    2. A problem in the sensor

        The driver enters the car and starts the Car. The On Board diagnostics diagnoses the car and found that there is some problem with one of the sensor. The Interface informs the problem to the driver with the message that the 'car needs maintenance'.

    3. Diagnosis Failure

        The driver enters the car and starts the car. The system interface notifies Diagnose failure. The system reboots the OBD in case of failure and reports errors if any.

    4. Problem in the software

        An error is detected in the software during startup. The system informs the user and tries to reboot once. If it does not solve the problem, automation is disabled until maintenance is carried out.

© The Editor(s) (if applicable) and The Author(s), under exclusive license
to Springer Nature Singapore Pte Ltd. 2021
M. Singh, *Information Security of Intelligent Vehicles Communication*,
Studies in Computational Intelligence 978,
https://doi.org/10.1007/978-981-16-2217-5

1. Car moving on the highway.

   ### 1. No problem

   The driver is moving on the highway towards his destination. The driver gets the traffic information, weather information from Wi-Fi devices on the intersections. The driver also gets the information of additional services like golf outlet, amusement park etc. with ongoing benefits (The discounts going on various services) via interface.

   2. Accident happens within the range of Ad-hoc network of the user.

   The system alerts the driver about the accident. System notifies the user verbally and visually, the details of the accident, like location of accident, time of accident and the current status of the traffic at the location of the accident from the intersection as well as from the cars in the range of ad-hoc network of the accident. The system also suggests alternate route to the destination in case of congestion on the current route.

   3. In front car of the user met with an accident

   The system automatically slows down the speed of the car and alerts other vehicles in close proximity of the user, about the accident. The system collects the accident information, like the pre and post-accident video, audio, pictures, location, time and sends the accident data captured to the nearby intersection device and to other cars in close proximity of the user and in the range of the ad-hoc network.

   4. The car itself met with an accident

   The system automatically captures all the information from the OBD i.e. status of all the sensors, the pre and post-accident pictures, video, audio clips, owner's car number, name, mobile no., address and other important details and sends it to the intersection device and other vehicles in the range of the ad-hoc network.

2. Data processing at the device on the intersection

   1. Non Critical data filtered out at the Intersections.

   The device at the intersection filters out the non-critical data and sends only the critical data (accident data) to the cloud server.

   2. Sending critical (accident) data to the authorities and injured person's family members.

   The server analyzes the data and sends the data in real time to authorities like hospital, police station, fire brigade (in case of fire) for investigation and also to the family members of the injured person in the accident.

Appendix B
Goal Model

1. *Notations Used in KAOS Model*

   For convenience we have broken down the KAOS model in smaller diagrams. We used different boxes for requirements, operation, agent, soft goal and different relationship between them as shown in Fig. B.1.

2. *Initialization Successful shows in* Fig. B.2.

3. Safe Drive to Destination presents in Fig. B.3.

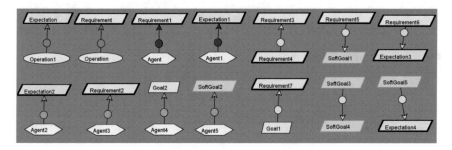

Fig. B.1 Show the notations used in the KAOS model

M. Singh, *Information Security of Intelligent Vehicles Communication*,
Studies in Computational Intelligence 978,
https://doi.org/10.1007/978-981-16-2217-5

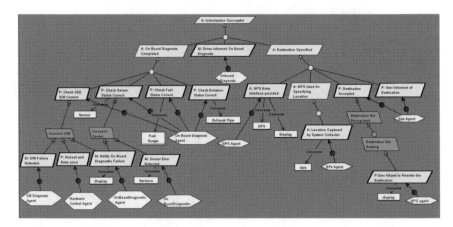

Fig. B.2 Successful initialization

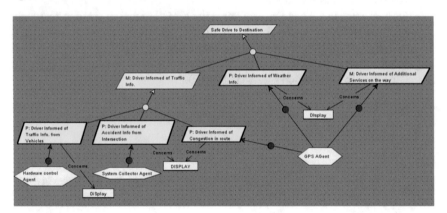

Fig. B.3 Safe drive to destination

Appendix C
Responsibility Diagrams

1. *On Board Diagnostic Agent as we can see in* Fig. C.1.
2. *Graphical Positioning System (GPS) Agent represents in* Fig. C.2.
3. *Law Management Agent has presents in* Fig. C.3.
4. *Hardware Control Agent shows in* Fig. C.4.
5. *Environment Context Agent depicted in* Fig. C.5.

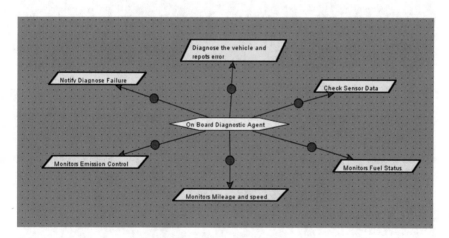

Fig. C.1 On board diagnostic agent

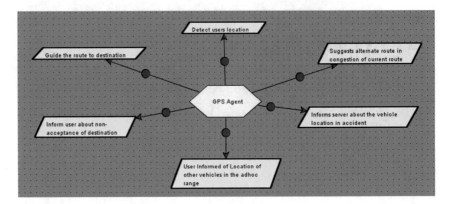

Fig. C.2 Geographical positioning system agent

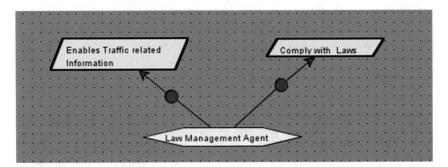

Fig. C.3 Law management agent

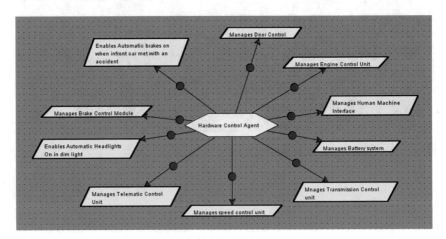

Fig. C.4 Hardware control agent

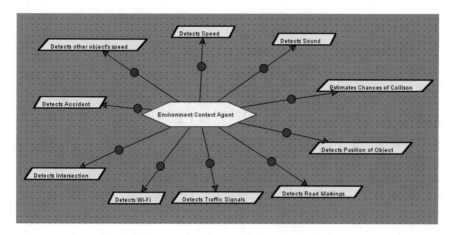

Fig. C.5 Environment context agent

Printed in the United States
by Baker & Taylor Publisher Services